Contents

KV-243-077

	Equipment and materials	v
	Introduction	vi
1	Collection and display	1
2	Averages	15
3	Correlation	28
4	Frequency distributions	41
5	Running totals	53
6	Probability	61
7	Sampling	71
8	Spread	81
9	Continuity	86
10	Further averages	91
11	Diagrams	97
12	Cumulative frequency and standard deviation	107
13	Exercises	119
	Answers	130

Acknowledgements

The author and publisher are grateful to the following for permission to reproduce copyright material: Mr E. H. Shephard and Methuen & Co Ltd for the drawing of Pooh Bear on page 71; Mr G. C. Dickinson and Edward Arnold Ltd for the tables from *Statistical Mapping* on pages 97, 101 (below), 102 (above) and 104 (below); Thomson Newspapers Ltd for the tables on pages 100, 101 (above) and 102 (below); Associated Lancashire Schools Examining Board (AL) for questions from the 1967 CSE statistics paper; East Anglian Examinations Board (EA) for questions from the 1966–7 CSE mathematics papers; Middlesex Regional Examining Board (MX) for questions from the 1966–7 CSE further mathematics (statistics) papers; South-East Regional Examinations Board (SE) for questions from Part 3 of the 1965–7 CSE statistics papers; Southern Regional Examinations Board (SR) for a question from the 1967 CSE 2B paper (where necessary questions have been metricated).

Grateful thanks are also given for the help received from G. A. G. Cox, C. V. Kingston, R. A. Parsons and R. W. Strong.

The illustrations are by David McKee and are based on original contributions by Ann Sinclair.

This book is due for return on or before the last date shown below.

STA
STA

K Lewis Co
H Ward We

Longman

LONGMAN GROUP LIMITED
London
Associated companies, branches and representatives throughout the world

© *Longman Group Ltd 1969*
First published 1969
SBN 582 20863 7

Made and printed in Great Britain by William Clowes and Sons Ltd, London and Beccles

Also in this series:
Home Mathematics by A B Tookey

Equipment and materials

Rulers, protractors, compasses
Graph and squared paper
Coloured card, coloured gummed paper
Balsa strip and cement
Scissors
Packs of playing cards
Dice: a set of 36 is desirable
Bathroom scales, height measurer
Potatoes or lino for linocuts; knives and paint
Maximum and minimum thermometers
Rice
Ball of string

Some tasks require special equipment, in particular see:
Permanent bar chart in chapter 1, page 8
Quincunx in chapter 4, page 51
Sampling in chapter 7: the bean experiments page 72: ball-bearing sampler
page 78: random electrical devices page 79.

It is suggested that folders be used for this work.
In many cases, group work is advisable.

Introduction

The authors believe that a knowledge of statistical ideas is of immense value for anyone living in the modern world and are convinced that statistics, a branch of mathematics so closely related to real situations, is a very teachable subject, particularly suited to discovery methods. It can often be related directly to the interests of the pupils themselves and for this reason it frequently appeals to pupils, particularly girls, who have found difficulty with a general mathematics course.

The inclusion of statistics as a topic in so many CSE papers adds weight to this belief. Throughout this book the authors have stressed the practical aspects of the subject. Students are encouraged to get their own facts, and to discuss, analyse and justify any conclusions they may deduce from the facts. The authors cannot emphasise too strongly their absolute conviction that the practical and experimental work in this course must not be neglected: indeed full benefit will not be gained if a 'do-it-yourself' approach is not adopted as a guiding principle.

Every effort has been made to cover the requirements of the many CSE syllabuses, but some topics (quota sampling, rank correlation, conditional probability and lengthy formulae for standard deviation) were felt to be too advanced for a suitable presentation at the CSE standard, and have not been included in this book.

Metric units and decimal currency have been used throughout.

1 Collection and display

Modern life continually demands decisions of us: what job do we want to have, what subjects shall we study at school, how shall we pass our spare time, how do we spend our money? Many of these questions can be answered quite simply, but many problems cannot be decided without further information.

In this first chapter, we will investigate some of the more common difficulties that arise when we collect facts or make a SURVEY, as it is called, and some of the useful ways in which the results of a survey can be displayed for other people to understand.

Let us pretend that we have been asked to carry out a survey to find out how many people live in the different types of house in the neighbourhood.

What sort of house do you live in? Is it detached, semi-detached, single storey, multi-storey, part of a terrace, one of a block of flats or none of these? Do you live in a caravan, a boat or an hotel?

Already, before we have even started counting how many people are in each group, there are many points that we will have to clear up. We must decide whether to distinguish between two- and three-storeyed houses; we must decide whether the end house of a terrace is the same as a semi-detached house (after all, each is joined to only one other house); we must decide how to distinguish between flats and terraces; we must decide how many of the 'special' houses must be mentioned separately, or do we expect to find so few that we can refer to them all as 'others'?

Class discussion

What would be sensible grouping of types of houses for your own class? Think of any awkward cases that might arise and decide what to do about them.

You will have already realised the most important thing about any survey: every house must be put into one group or another. Everyone must understand exactly what the group headings mean.

Let us suppose that we have decided to use the following group headings: detached multi-storey houses, detached single-storey houses, semi-detached houses, terraced houses, flats and other houses (to allow for the people who live in a barn).

We are now ready to start our survey. We take our notebook and pencil, and decide to stand outside the Town Hall on Saturday morning, and ask people we see to tell us the type of house they live in.

On the next page is the result of the survey.

Detached		Semi-detached	Terraced	Flats	Others
Multi-storey	Single-storey				
6	5	10	5	9	1

(The 'other' entry happened to be the caretaker of a factory who had a bed-sitting-room on the premises.)

We must now decide how to present this information. This will depend to a very large extent on who wants to use the facts we have obtained. The local housing authority would probably be quite happy with the way we have already laid it out: the information is clear, and is easy to store for future reference. Such a presentation is called a TABLE or a TABULATION.

A newspaper might record the information like this:

Detached	multi-storey	𝕩 𝕩 𝕩 𝕩 𝕩 𝕩
	single-storey	𝕩 𝕩 𝕩 𝕩 𝕩
Semi-detached		𝕩 𝕩 𝕩 𝕩 𝕩 𝕩 𝕩 𝕩 𝕩 𝕩
Terraced		𝕩 𝕩 𝕩 𝕩 𝕩
Flats		𝕩 𝕩 𝕩 𝕩 𝕩 𝕩 𝕩 𝕩 𝕩
Others		𝕩

This is called a PICTOGRAM and is frequently used in posters and advertising displays when it is important to attract attention.

See if you can discover any examples of pictograms in your neighbourhood. You will probably find some in the Post Office, the Citizens' Advice Bureau, the Town Hall or in any local government office.

Another very common way of presenting numerical information in a see-at-a-glance form is the BAR CHART.

SURVEY OF TYPES OF HOUSE IN ANYTOWN
Number of people questioned: 36. Date: 29 February 1969

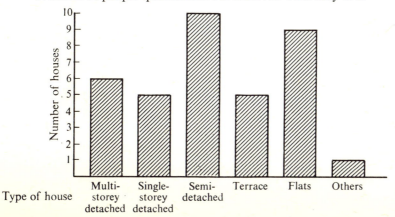

Alternatively, a bar chart can be drawn sideways:

SURVEY OF TYPES OF HOUSE IN ANYTOWN
Number of people questioned: 36. Date: 29 February 1969

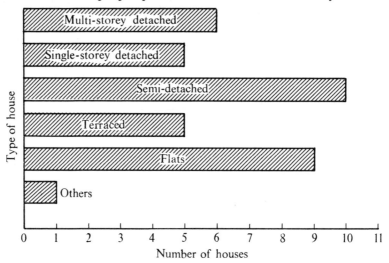

A bar chart should always have a heading, describing exactly what information is illustrated, with the date of the survey. The scale used to represent the numbers should be stated, and each column or bar should be labelled. Notice that it is simpler to do such writing on the sideways chart. You will find that sometimes the gaps between separate bars are left out; this saves space but it is not so easy to distinguish the different groups if this is done.

Does the order in which the bars are placed matter?

Exercise A

1. Carry out a survey to find out the sorts of houses that the people in your form live in. Present your answers as a table, a pictogram and a bar chart.

2. If another class carried out a similar survey, would you expect any great difference between the results? If this other class was half the size of your class, what would you expect to happen? Would the same overall pattern be obtained by a class in a different part of the country? Draw what you think might be the chart you would obtain from a school in an English industrial city, from a school in a Welsh mining village, from a school in the Scottish Highlands.

3. Decide on the possible group headings that would be necessary for a survey of pets owned by members of the class. Carry out such a survey and present the information in a suitable form.

4. Make a list of the subjects that you study at school. Can any be grouped together under the heading 'Others'? Work out the number of minutes devoted to each subject in a week and display your results as a bar chart.

5. Decide on some six to eight headings into which your day could be divided, such as travelling to school, eating, sleeping, watching TV, etc. Draw bar charts to illustrate the difference between a school-day and a non-school-day. Does it matter which school-day you pick?

6. Carry out a survey to discover how your class travels to school each day. The usual preliminary discussion will be needed and you will also have to agree what to do with such difficult cases as the person who walks part of the way and travels by bus for the rest.

Pie charts

Another way of displaying the results of a survey is by use of a PIE CHART, in which a circle is cut up into slices or sectors of different sizes.

In the type-of-house survey already discussed, there were 36 people questioned altogether. At the centre of a circle there are 360°, which means that in this particular survey, each person can have a $360° \div 36 = 10°$ sector of the whole pie.

The 6 who live in multi-storey detached houses will therefore altogether need a sector measuring 60°.

The angular measurements for all groups will be given by the table:

| | Detached | | | | | | |
	Multi-storey	Single-storey	Semi-detached	Terrace	Flats	Others	Total
Number of people	6	5	10	5	9	1	36
Number of degrees in pie chart	60	50	100	50	90	10	360

We now use a protractor to measure these angles at the centre of a circle. A suitable radius is usually 5 or 6 cm for an exercise book, but should be much more if the chart is to be displayed on a wall.

TYPES OF HOUSE IN ANYTOWN
Number of people questioned: 36. Date: 29 February 1969

10° represents 1 person

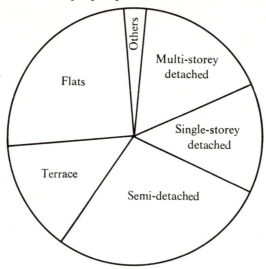

Measure the angles at the centre for each sector and check that this chart agrees with the table above.

Does the order of the sectors matter?

It is fairly easy to determine from a pie chart the relative sizes of each group: the big sectors can be readily distinguished from the small ones and it is easy to judge by eye what fraction each one represents. This is often the most important aspect of a survey. A pie chart is a very concise way of presenting information; if there are not many groups, quite a small circle gives the information clearly, and with a protractor the actual numbers can be obtained fairly accurately.

Exercise B

Calculate the number of degrees required for each sector to represent these tabulations as pie charts. Construct the charts.

1. Survey of left- and righthandedness in a class of boys and girls.

	Righthanded		Lefthanded		Total
	Boys	*Girls*	*Boys*	*Girls*	
Numbers	16	9	6	5	

2. Survey of number of brothers and sisters for members of a class.

Number of brothers and sisters	0	1	2	3 or more	Total
Number of pupils having that number of brothers and sisters	6	8	3	1	

3. Survey of absentees from a particular school during one week.

Number of times absent in the week	1	2	3	4	5	Total
Number of pupils absent that number of times	17	25	15	10	5	

Unfortunately, if the total number of cases is not a good number like 36, 18 or 72, the calculation of the exact number of degrees for each sector is not always simple; this is one disadvantage of a pie chart. In practice, of course, a slide rule can be used for such work and if you already know how to use this instrument, the calculations for a pie chart should present no difficulty.

Class projects

1. Make a class pie chart. Calculate the angle of the sector for each individual member of your class by dividing 360° by the number of pupils in the class.

Decide on a suitable radius and cut out a supply of individual slices from coloured gummed paper.

Prepare a baseboard and construct the pie chart of the survey carried out concerning the types of houses for your class. The same idea can be used for other class surveys.

If you write your name on your sector, the final chart contains more infor- mation than usual as it is also clear which pupils are in each group. Make sure that each group is composed of sectors which are all the same colour.

What will have to happen if the number of children in your class changes? What will you do if somebody is absent for a particular survey?

2. Make a class bar chart. Cut out small squares of coloured gummed paper and write your name on your squares. Use a large sheet of stiff paper or card and construct bar charts of the class surveys you do, using the coloured squares to make the columns.

What will happen if the number of pupils in the class changes?

Will it matter if some people are absent for a particular survey?

3. Make a permanent bar chart. A permanent bar chart can be made from pieces of hardboard, which can be placed between narrow strips of wood fastened to a baseboard, these can be T section or built up from two strips of different widths. A look at the mouldings in a do-it-yourself shop will help. If the baseboard is painted with blackboard paint, titles and dates can be written on with chalk and easily changed.

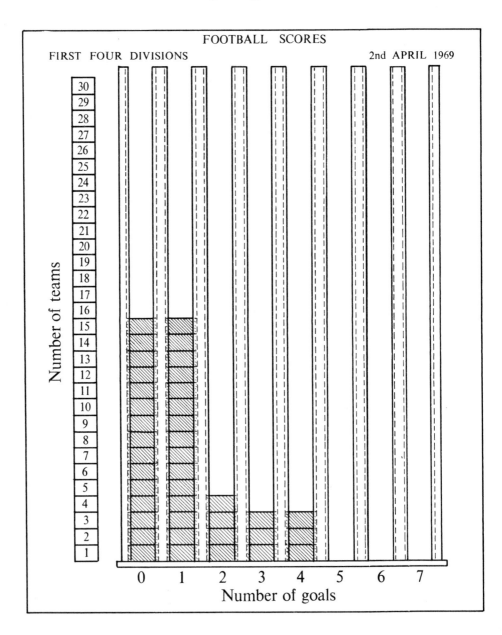

FOOTBALL SCORES

FIRST FOUR DIVISIONS   2nd APRIL 1969

Number of teams

30
29
28
27
26
25
24
23
22
21
20
19
18
17
16
15
14
13
12
11
10
9
8
7
6
5
4
3
2
1

0   1   2   3   4   5   6   7

Number of goals

Such a bar-chart kit can be used for any survey. One suggestion is to record week by week the number of goals scored by the teams in the first four divisions of the Football League. Two weeks gave the number of teams with particular goal scores as follows:

Number of goals	0	1	2	3	4	5	6	7
Number of teams 18 March	29	30	17	6	1	1	—	—
Number of teams 25 March	25	23	23	7	9	1	—	—

A small group of pupils could be selected to keep such a record up to date, and would be responsible for the changes each Monday.

4. A vehicle survey. A survey which your class can actually carry out is the investigation of traffic conditions on a road near your school. Even twenty minutes is long enough on a reasonably busy road for a survey to be very worth while.

Such a survey is exactly the sort of investigation that would be carried out if there was a proposal to have a new pedestrian crossing, or to make a one-way street, or to declare it a clearway.

A preliminary discussion will be essential, and points which must be decided would include:

a. What is the best time of day for such a survey?

b. Will one survey be sufficient, or should we return on a second, or even third occasion?

c. What other factors will affect the results? Will rain, snow or fog give a false impression?

d. Shall we count the traffic in one direction only, or in both directions?

e. How shall we classify the various types of vehicle?

f. Will it help to have a special form to record the information? If so, what should such a form look like?

g. Ought we to have a trial run to see if there are any other difficulties?

h. Which pupils are going to be allocated to which jobs? Do we need a chief organiser, and if so, who? Ought we to work in pairs in order to check the actual results?

There will be many other questions which your particular choice of road will suggest, but once these are settled, the actual counting is quite an easy task.

A suggested form which could easily be duplicated or copied is shown on the next page:

Vehicle survey

Name....................... Form............. Date.............
Location ....
Direction of traffic..
Time started................. Time finished.................

Type of vehicle	Tally	Total
Private cars and vans		
Lorries and commercial vehicles		
Public transport		
Two-wheeled motor vehicles		
Pedal cycles		
Other types		
Remarks	Grand total	

These classifications might lead to difficulties so you will have to decide about awkward cases which arise when you get back to the classroom.

You will see that a space has been left on the form headed 'Tally'. It is in this space that a mark can be made to record each vehicle as it passes; the marks are totalized once the actual survey is completed.

In practice, to simplify the final counting, a tally is usually recorded by the four-bar gate method in which every fifth mark is used to link together the previous four marks. Thus the final situation might be:

  ‖‖‖ ‖‖‖ ‖‖‖ ‖‖‖ ‖‖‖ ‖‖‖ ‖‖‖ ‖‖ rather than:

‖‖‖‖‖‖‖‖‖‖‖‖‖‖‖‖‖‖‖‖‖‖‖‖‖‖‖‖‖‖‖‖‖‖‖‖‖‖

It is much simpler to total in the first case than in the second; try it and see.

To complete the survey the results must be presented in a clear manner. It would be a good idea if tables, bar charts, pie charts and even pictograms were used.

5. Other subjects suitable for class surveys

 a. Keep a record of school menus for the week and on Friday have a vote to decide on the order of popularity. Will you allow one vote for each pupil?

What about two meals which each receive the same number of votes?

Rank and display your results in the manner you think is most appropriate.

b. Make a survey of the number of books borrowed each day from the school library and display a week's results. Try again in other weeks and see if there is a pattern.

c. Decide amongst yourselves a choice of six weekly TV programmes and carry out a vote to discover the order of popularity.

d. What is the most common length of sentence, in words, in a particular book?

Does sentence length vary greatly from author to author?

Would sentences in foreign books have a different length pattern?

e. What is the most common length of English word?

Does word length vary from author to author?

Does German have the same word-lengths as English?

Exercise C

1. A stamp collector is given a packet of mixed stamps as a birthday present. The stamps come from the following countries: Australia, Belgium, Canada, Denmark, Egypt, France and Germany.

On spreading the stamps out on a table he found they were:

```
A  G  F  C  E  A  A  A  G  D  G  C
F  F  F  D  B  C  A  A  G  F  G  F
F  D  A  A  A  A  F  F  F  F  F  B
E  G  G  F  A  A  G  C  G  A  D  C
G  B  C  G  F  A  A  F  A  A  C  D
```

(A represents an Australian stamp, B a Belgian stamp, C a Canadian stamp and so on.)

Put these results into a table, display as a bar chart and as a pie chart.

2.

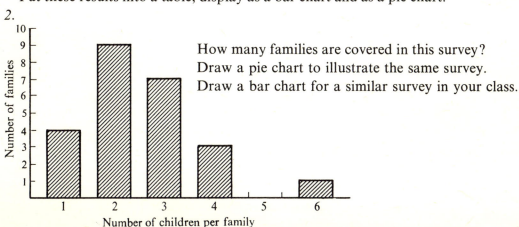

How many families are covered in this survey?
Draw a pie chart to illustrate the same survey.
Draw a bar chart for a similar survey in your class.

3. In a certain newspaper round, the following numbers of papers are delivered:

The Semaphore 3, *Moon* 25, *Rapid* 9, *The Glance* 15 and *Reporter* 8.

Draw a pie chart to illustrate this delivery.

The *Reporter* is forced to close down, and the eight regular readers of that paper all decide to take the *Rapid* instead. Draw a pie chart for the new delivery pattern.

4. The following table is the result of a popularity vote on school meals during a certain week:

Fish and chips	Stew	Cold meat	Pie	Sausages
12	3	2	8	5

Illustrate this table as a bar chart and as a pie chart. Compare these figures with the results obtained from your own survey.

5. During a fishing holiday a boy caught 2 pike, 6 bream, 10 roach and 12 perch. Make a potato- or linocut of a fish and use it to make a pictogram displaying this information.

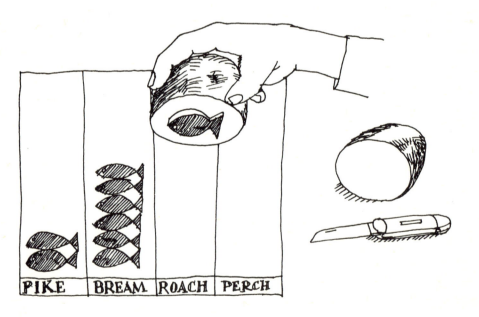

6. The pie chart illustrates the results of a survey carried out in a class of 36 girls. How many are there in each category?

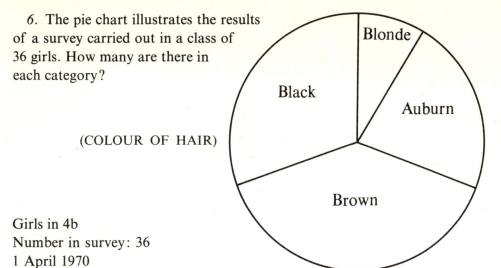

(COLOUR OF HAIR)

Girls in 4b
Number in survey: 36
1 April 1970

7. A record of TV broadcasting for a month gave the following number of hours:

Sport	19	Feature films	17
Westerns	9	Pop	20
Adventure	4	Serials	26
Travel	3	Education	12
Quiz games	9	News	12
Variety shows	18	Plays	20
Science fiction	11		

a. Illustrate this in a pie chart.

b. Reduce the number of categories by suitable regrouping so that there are no more than eight. Draw a pie chart of this new table and compare it with the one you drew in answer to (*a*). Discuss the desirable number of categories if pie-chart illustration is to be used.

8. See if you can obtain a sheet of Letraset transfers or a printing outfit; see how many of each letter is included. Are there as many Es as Zs? Imagine you are going to start a factory to produce alphabet transfers, and devise an experimental survey to decide the relative numbers of each letter you are going to produce.

Who else might find the result of your survey of interest?

Would your results apply equally well if your factory was producing for a foreign market, say France or Germany?

9. Here is a very good example of the display of information, but with one important exception. Can you see what it is?

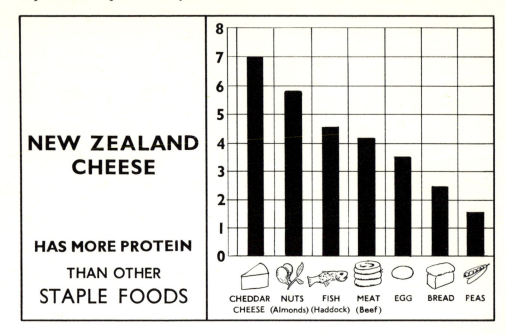

NEW ZEALAND
CHEESE

HAS MORE PROTEIN

THAN OTHER

STAPLE FOODS

CHEDDAR NUTS   FISH   MEAT   EGG   BREAD PEAS
CHEESE (Almonds) (Haddock) (Beef)

2 Averages

The average family is composed of 1 underpaid male, 1 overworked female and 2·2 underfed children

The word 'average' is used a lot; we often hear of the average man, above average marks, on the average, and so on. 'Average' is used here in the sense of 'typical', 'usual' or 'normal'. We all know how to find the average of a set of numbers: add them all up and divide by how many numbers there are.

To find the average of 4, 6 and 11 first add up, which gives 21, and then divide by 3 because there are three numbers:

$$\frac{4+6+11}{3} = \frac{21}{3} = 7.$$

The average of 4, 6 and 11 is 7.

This is really only one special sort of average. In this chapter we shall see that there are other sorts of average and we must be clear which is which when we use the word.

The shopkeeper's average

A man who owns a shoe shop checks his day's sales and finds he has sold shoes of sizes:

3, 6, 4, 7, 7, 8, 4, 7, 8, 7.

If he calculated the average size shoe sold he would get:

$$\frac{3+6+4+7+7+8+4+7+8+7}{10} = \frac{61}{10} = 6\cdot1$$

which is difficult to understand since shoes of size 6·1 do not exist.

However, what is important is that more size 7 shoes were sold than any other size and this will be the average as far as the shopkeeper is concerned.

The correct name for this sort of average is MODE.

The mode does not even have to be a number. For example, if ten girls were asked to name their favourite colour and their replies were red, blue, blue, yellow, red, red, green, red, red, orange, then the mode would be red.

The *mode* is the most fashionable item, the most popular item, the one that occurs most frequently.

Example

Find the mode of the following:

      0, 5, 4, 5, 0, 1, 4, 3, 3, 4, 2, 1, 5, 4, 5, 5
      0, 1, 2, 2, 3, 5, 3, 5, 1, 5, 4, 4, 4, 3, 5, 5, 1.

In this case it would be wise to set out the numbers 0 to 5 and make a tally:

Number	Tally	Total
0					3
1	⊥⊦⊦⊤	5
2					3
3	⊥⊦⊦⊤	5
4	⊥⊦⊦⊤ ||	7
5	⊥⊦⊦⊤ ⊥⊦⊦⊤	10

and from this we can see that the mode is 5.

There is no reason why there should not be more than one mode. The man in the shoe shop might check his sales on another day and find:

      6, 4, 4, 5, 7, 8, 5, 8, 8, 9, 8, 5, 5, 4, 5, 8, 6, 7.

In this case five pairs of size 5 were sold and also five pairs of size 8 were sold. Both proved to be equally popular, the modes being 5 and 8. This is called a BI-MODAL situation.

Exercise A

   1. Find out the favourite colour in your class.

   2. Find the modal size of shoe in your class.

   3. Find the modal way of travelling to school in your class.

   4. Find the mode of these numbers:

      2, 3, 3, 3, 4, 2, 6, 1, 5, 1, 1, 4, 4, 3, 5, 2, 3, 2, 1, 1, 1, 3, 1, 5, 4, 5, 1,
      2, 2, 2, 3, 4, 7.

5. Tabulate and tally this collection of numbers and find the mode.

1, 4, 7, 4, 3, 8, 0, 0, 1, 8, 2, 9, 2, 6, 0, 8, 2, 0, 6, 2, 8, 1, 9, 0, 3, 7, 1,
0, 7, 5, 1, 9, 8, 3, 2, 6, 2, 0, 6, 2, 5, 2, 8, 7, 3, 0, 1, 2, 2, 8.

6. Describe a situation where every item is the mode.

7. If the modal height of ten boys is 150 cm, what, if anything can be deduced about the heights of this group?

The one in the middle
Here is the school netball team:

Line them up by height from the shortest to the tallest and we get:

| 1 | 2 | 3 | 4 | 5 | 6 | 7 |

Girl Number 4 occupies a very special position; she is in the middle, there are as many girls shorter than her as there are girls taller than her. As far as height is concerned she is in the MEDIAN position and her height is the MEDIAN HEIGHT of the team.

The median is another sort of average.

To find a median it must be possible to arrange items in order or to rank them in positions, first, second, third and so on. Once this is done, you simply take the one in the middle as the median.

To find the median weight of the six boys in the boxing team we would first weigh them and then arrange them in order from the lightest to the heaviest:

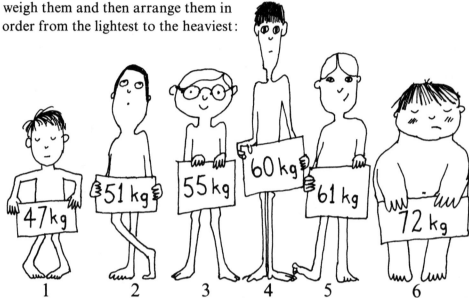

Now when you try to find the one in the middle a difficulty arises because there is nobody in the middle; another way of looking at it is to say that boys 3 and 4 share the middle place and between them make the median weight.

Boy Number 3 weighs 55 kg and boy Number 4 weighs 60 kg, the median weight is therefore $(55 + 60) \div 2$; that is, 57·5 kg.

Notice that in this case it is not the weight of any actual boy in the team.

Example

Here are the marks out of ten given to a class in a test. Find the median mark.

5, 9, 8, 5, 6, 7, 8, 4, 3, 9, 7, 7, 8, 6, 6, 5, 2, 10, 5, 4, 8, 9, 6, 4, 3, 8, 9.

First of all rewrite the list in order from lowest to highest.

2, 3, 3, 4, 4, 4, 5, 5, 5, 5, 6, 6, 6, 6, 7, 7, 7, 8, 8, 8, 8, 8, 9, 9, 9, 9, 10.

Now count how many entries have been made—27—and the median mark will be half-way, the 14th. Count along 14 places and the mark occupying this position is 6.

Exercise B

1. Line up the class by height. Who has the median height? What is the median height? Compare the median height for the girls with the median height for the boys.

2. Make a collection of bottles and arrange them by height to display the median.

3. Here are the marks out of 10 given to a class in a test; find the median mark:

8, 7, 4, 10, 1, 5, 6, 6, 5, 4, 3, 4, 8, 7, 10, 4, 5, 3, 2, 9, 7.

4. Try again with these marks:

9, 9, 7, 6, 7, 4, 3, 2, 3, 7, 7, 6, 5, 7, 5, 8.

5. Here are the percentage marks obtained in a mathematics exam:

70, 72, 30, 74, 80, 83, 36, 50, 38, 84, 38, 85, 92, 50, 70, 68, 17, 48, 77, 72, 60, 74, 14, 75, 83, 65, 33, 52, 46, 34, 32.

Find the median. Next time you have an examination do not just concern yourself with your position in class, have a look and see if you are above or below the median.

6. See if you can find the median mark from this table without writing them out in order:

Mark	Number of pupils
10	1
9	4
8	2
7	6
6	8
5	7
4	3
3	1
2	2
1	0
0	1

7. Here are the marks out of 10 in three tests with the same pupils. Find the median mark for each test.

Mark	Number of pupils		
	Test 1	Test 2	Test 3
10	1	2	1
9	3	5	2
8	0	4	2
7	5	6	6
6	7	8	5
5	8	4	3
4	5	3	4
3	1	1	4
2	3	0	1
1	0	0	2
0	0	0	2

The ordinary average

The ordinary average is the one we were talking about at the beginning of this chapter where you add the numbers up and divide by how many there are. The proper name for this sort of average is the ARITHMETIC MEAN: for the rest of this book we will often refer to it as just the MEAN.

Here is a flow diagram showing the way a computer would be programmed to find the mean of a set of numbers:

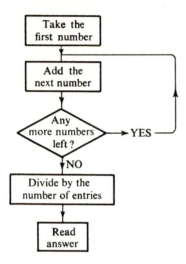

Exercise C

1. Find the mean of: 4, 7, 8, 5, 4, 3, 2, 9, 8, 6.

2. A cricketer scores 21, 40, 8, 73, 68 on five completed innings, find his mean score.

3. If the cricketer in question 2 went in for another innings and scored 0, what would his average be then?

4. Find out what time everybody in your class went to bed last night (work to the nearest half-hour). Find the mean of these answers and state how much above or below the mean your time was.

5. A car covers a journey of 315 km in 7 hours. What is the mean speed?

6. If a car travels at an average speed of 55 km/h for 4 hours how far does it go?

7. The average of 10 numbers is 8·6. What is the total of the 10 numbers?

8. If the average of 4 numbers is 8 and the average of another 6 numbers is 14, find the average of the combined set of 10 numbers.

9. The average of 3 numbers is 2·7 and the average of another 7 numbers is 8·6; find the average of the combined set of 10 numbers.

Variation from the mean

The mean of a set of numbers can be portrayed graphically. Consider this set of scores: 7, 2, 6, 4, 6.

First of all represent these on a bar chart.

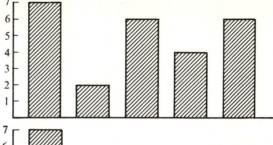

Work out the mean, which is 5, and draw a line through your chart to show where it comes.

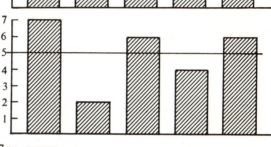

Chop off all the pieces above the line; they will fit in below the line and fill the gaps up.

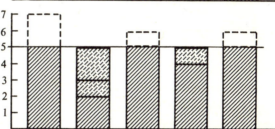

The total of the pieces above the average line will always be the same as the total of the gaps below the average line.

This fact can also be dealt with numerically. Consider the same numbers, 7, 2, 6, 4, 6. We shall write them out in a column and see how much each one differs from the mean of 5. The sum of all these differences should be zero.

	Score		Difference from 5	
	7		+2	
	2			−3
	6		+1	
	4			−1
	6		+1	
Total	25	*Total*	+4	−4 = 0
Mean	5			

(Notice that this calculation is simplified by putting positive differences in one column and negative differences in a second column.)

Another point worth noting is that the mean always lies between the greatest and the smallest of the numbers you are considering. For example, if you had to find the mean of 3, 4, 5, 8, 12 and it came to 16·4 you would know that an error had occurred. What is the real answer in this case? What mistake was made in the calculation?

Exercise D

1. Represent these scores on a bar chart: 5, 8, 2, 4, 7, 10.
Draw a line through the position of the mean. Make another chart showing how the pieces above the line fit into the spaces below the line.

2. Make a model to illustrate question 1. Use a baseboard and cardboard bars which are built up from unit pieces so that they can be moved about. For example:

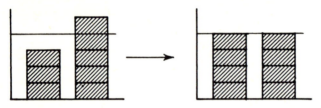

An even better model could be made by piling up blocks or bricks against a backboard.

3. Copy and complete this table to show that the sum of differences from the mean is zero.

Score	Difference from mean	
	+	−
27		
29		
46		
18		
54		
60		
Totals		
Mean		

4. Make a table for the following numbers and carry out the same sort of calculation as in question 3.
8·6, 4·2, 2·9, 5·8, 2·3, 8·0, 7·4, 6·4.

5. State the limits between which the mean of these numbers lies:
159, 153, 160, 111, 172, 161, 159, 110, 108, 147, 171, 144, 161, 104, 128, 165.

We have seen that the mean always lies between the smallest and the greatest. Consider these numbers:

103, 102, 106, 107, 105, 101.

We can write these as:

$100+3$, $100+2$, $100+6$, $100+7$, $100+5$, $100+1$.

On further examination you will see that the mean must be:

$100+$(the average of 3, 2, 6, 7, 5, 1)

that is:

$$100+\frac{3+2+6+7+5+1}{6} = 100+\frac{24}{6},$$

which gives the mean as 104.

In certain cases where the numbers being considered are all close to one another we can save a lot of tedious calculation by this method.

The 100 in this example is called the ASSUMED MEAN; the 104 is the ACCURATE MEAN.

Here are two more examples.

1. Calculate the arithmetic mean of:

980·7, 981·4, 981·2, 980·6, 980·5, 981·6, 981·5, 981·0, 980·9, 981·7

Notice that every number is nine-hundred and eighty something. Take out 980 and average the pieces left over.

$$\text{Mean} = 980+\frac{0·7+1·4+1·2+0·6+0·5+1·6+1·5+1·0+0·9+1·7}{10}$$

$$= 980+\frac{11·1}{10}$$

$$= 980+1·11$$

$$= 981·11$$

The assumed mean is 980 and the accurate mean is 981·11.

Depending on what these numbers measure, we will often take the mean in such a case to be 981·1.

2. Calculate the mean of: 104, 110, 97, 106, 98, 101.

$$\text{Mean} = 100+\frac{4+10-3+6-2+1}{6}$$

$$= 100+\frac{16}{6}$$

$$= 100+2\tfrac{4}{6}.$$

The assumed mean is 100, the accurate mean is $102\tfrac{2}{3}$.

Time can also be saved with the following example. Calculate the mean of: 100, 400, 700, 1600, 200. Work in hundreds and the calculation becomes:

$$\frac{1+4+7+16+2}{5} = \frac{30}{5} = 6.$$

Answer: Mean = 600.

Upsetting the average

In the little village of Statterton live these people; their annual incomes are given:

	Annual income in £
The Squire	10,000
The Bailiff	950
Innkeeper	900
Postmaster	800
John ⎫	650
Mike ⎪	600
Eddie ⎬ Farm	600
Rod ⎪ labourers	550
Jan ⎪	550
Dick ⎭	400
10 people	*Total earnings* 16,000

Hence it is true to say that the mean earnings in the village amount to £16,000 ÷ 10, that is £1600 per head.

This is a very misleading picture because only one person earns more than this and all the others earn a great deal less. A more accurate picture would be obtained if the squire was left out of the calculation altogether.

Exercise E

1. Find the mean of:

43, 47, 42, 44, 43, 49, 48, 50

by taking out 40 from each number at the start.

2. Repeat question 1 but this time take out 45 to start with.

3. Does it matter which number you take out to start with? Try to explain your answer.

4. Use a short method to find the mean of:

106, 109, 110, 105, 107, 108, 106, 107, 109, 105.

5. Find the mean of:

1000, 6000, 8000, 12,000, 3000 by working in thousands.

6. Ten boys measured the length of a metal bar and obtained these results (in centimetres):

4·15, 4·12, 3·95, 4·00, 4·10, 4·08, 4·10, 4·00, 4·10, 4·05.

Find the mean length.

7. What would you do if the results for question 6 were given as:

4·10, 3·90, 4·00, 4·05, 4·00, 6·30, 4·10, 4·05, 4·14, 4·20?

Give an answer for the mean length of the metal bar in this case.

Exercise F

1. If a teacher marks an exercise out of 10 and then decides to add 2 to every mark, this diagram will represent the situation:

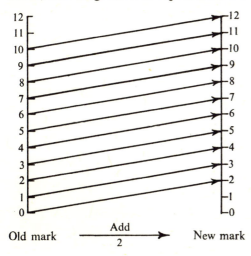

If the mean mark on the old scale was 7·4, what will be the mean mark on the new scale?

Draw a similar diagram to represent the doubling of every mark.

2. A teacher gave a group of 12 pupils the following marks:

44, 59, 62, 40, 80, 47, 68, 62, 25, 73, 60, 52.

a. Find the mean mark.

b. What happens to the mean if the teacher decides to add an extra five marks to every score?

3. Here are the marks given to a group of 14 pupils:

12, 9, 26, 14, 37, 14, 17, 21, 20, 32, 41, 29, 21, 17.

a. Find the mean mark.

b. What happens to the mean if the teacher decides to double every mark?

4. An experiment in science involved the accurate weighing of a substance; 15 boys recorded the weight (in grams) as follows:

43·20, 43·28, 43·29, 43·31, 43·40, 43·35, 43·30, 43·36, 43·32, 43·30, 43·32, 43·25, 43·28, 43·29, 43·29.

Find the mode, median and mean value.

Which of these three values would you take to represent the weight?

5. A darts player records these 12 results:

7, 12, 9, 4, 1, 11, 19, 17, 60, 20, 2, 6.

Find the median and the mean score. What about the mode?

6. A survey was taken with a form of 30 pupils to find how many children there were in their families (including themselves). The results were:

Children	Families
1	7
2	11
3	6
4	3
5	2
6	1

What is the best sort of average to use here?

7. Criticise the drawing and quotation at the beginning of this chapter.

8. Conduct in your own class a similar survey to that in question 6 and record the results. See if you can compare them with those of another class.

9. The staff in a factory consists of:

	Earnings per annum (£)
1 Owner	12,000
1 Manager	2,000
2 Foremen	900 each
10 Skilled operators	700 each
4 Unskilled workers	400 each

a. Find a mean annual wage and make the picture look as good as possible from the owner's point of view.

b. Find another mean annual wage to show how poorly paid the factory employees are.

10. If a train can travel at 120 km/h:

a. How far can it travel in 5 hours?

b. How long will it take to travel 300 km?

11. A car is used for a journey of 450 km: for the first 150 km the average speed is 75 km/h and for the remainder of the journey the average speed is 60 km/h.

a. How long did the first stage take?

b. How long did the second stage take?

c. Calculate the average speed for the whole journey.

12. The heights of pupils in a class were measured as (centimetres):

124, 124, 126, 130, 128, 122, 126, 116, 122, 140, 128, 130, 124, 118, 126,

124, 128, 130, 132, 118, 120, 136, 132, 130, 134, 128, 124, 128, 132, 120.

Work from an assumed mean of 124 and calculate the accurate mean. Why do you think 124 was suggested? Will you get a different result if some other assumed mean is chosen?

13. This bar chart represents a survey concerning colour of hair in a class of pupils:

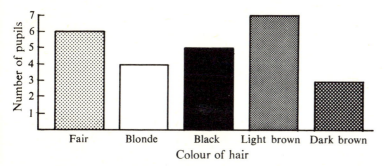

a. Give an average (a typical value) to represent this situation.

b. Carry out such a survey in your own class. Illustrate your results and give an average.

14. On what day of the week is your birthday this year? Conduct a survey in your class to find the answer to this question for everybody. Display your results in as many different ways as you can and give an average.

15. In a school athletics championships the following times in seconds were recorded for the 100 metres:

11·7, 11·8, 11·4, 11·5, 11·2, 11·9, 11·6, 11·5, 11·7, 12·0, 11·8, 11·3, 11·5,

11·7, 11·4, 11·7, 11·5, 11·8, 11·6, 11·4, 11·5, 11·3, 11·7, 11·6.

Find the mode, median and mean.

16. Find the mean of the following set of numbers and show that the sum of the differences from it is zero:

17, 56, 45, 68, 20, 46, 33, 51, 109, 7.

3 Correlation

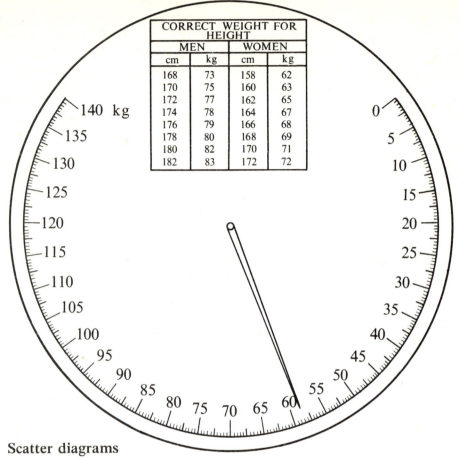

CORRECT WEIGHT FOR HEIGHT			
MEN		WOMEN	
cm	kg	cm	kg
168	73	158	62
170	75	160	63
172	77	162	65
174	78	164	67
176	79	166	68
178	80	168	69
180	82	170	71
182	83	172	72

Scatter diagrams

The chart on the weighing machine suggests that there is a connection between height and weight. To investigate this, the following experiment was conducted with 12 boys.

Everyone was weighed and measured, and here are the results:

Pupil	a	b	c	d	e	f	g	h	i	j	k	l
Weight (kg)	52	59	58·5	63	60·5	66	66	66·5	72	76·5	71	79·5
Height (cm)	122	124	126	127	128	128	132	134	136	139	142	144

These results can be plotted on a graph, taking the weight as the x number and the height as the y number for each pupil.

For example: for pupil *a* plot the point (52, 122)

　　　　　for pupil *b* plot the point (59, 124) and so on.

This gives:

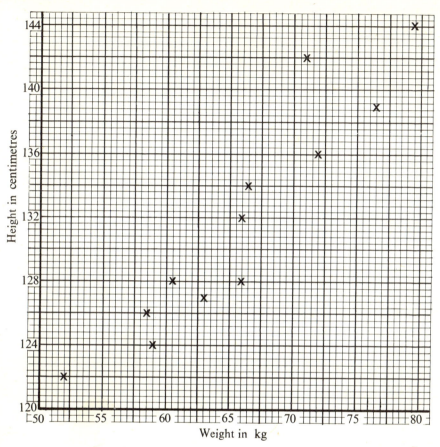

You will notice that these points are not scattered all over the place, but are confined to quite a narrow band:

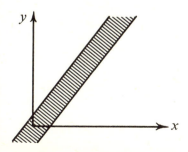

When this happens we say that there may be a correlation between the two sets of results.

Line of best fit

Instead of a band we draw a line which best represents this situation.

Place your ruler on the diagram and judge by eye a line which seems to go right through the middle of the band. If you do this it is best to use a transparent ruler so that you can see all the points as you move the ruler about.

A more accurate method is to calculate and plot the mean for each set of measurements, e.g. the mean weight is:

$$\frac{52+59+58\cdot5+63+60\cdot5+66+66+66\cdot5+72+76\cdot5+71+79\cdot5}{12}$$

which gives 65·875 kg.

By the same method, the mean height comes to 132·67 cm.

Put a ring round this point (65·875, 132·67) so that you do not mistake it for one of the actual set of marks. Make sure that your line of best fit passes through this point by placing your ruler against it and regarding it as a pivot until you have decided which line to draw.

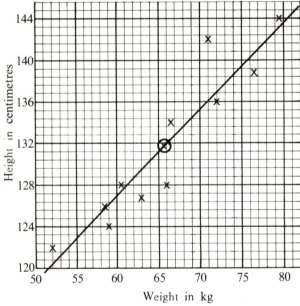

Use of correlation

When a line of best fit can be drawn and it slopes upwards from left to right:

we say that a direct correlation probably exists between the two sets of values.

This tells us that a high value along one axis corresponds to a high value on the other, and also that a low value corresponds to a low value.

The investigation about height and weight suggests that a tall person is heavy and a short person is light. The scatter diagram can be particularly useful to help estimate the weight of a person if we only know the height, and vice versa.

Example

If we knew that a pupil weighed 65 kg, we can make an estimate of his height as 131 cm.

A pupil who was 138 cm high could be estimated as weighing 73·5 kg.

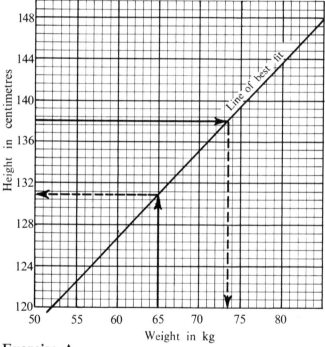

Exercise A

1. This table gives the marks scored by a group of 10 pupils in a French test and in a German test. Plot the results on a scatter diagram, find the mean of each test and plot a point representing the two means. Are we justified in assuming a direct correlation between ability in French and ability in German? Draw a line of best fit.

Pupil	a	b	c	d	e	f	g	h	i	j
French	42	32	55	50	67	73	68	77	19	43
German	42	44	52	51	53	66	56	60	18	37

2. Use your graph from question 1 to answer the following questions:

a. If a pupil was absent for the French test but obtained 32 in German, suggest a suitable mark.

b. Estimate a German mark for a pupil who received 71 in French.

3. The following pairs of values of x and y were obtained in an experiment:

x	0·6	1·1	1·7	2·2	2·7	3·2	3·8	4·7
y	2·2	2·4	2·7	3·2	3·3	3·5	4·0	4·5

You are told that there is reason to think that direct correlation exists so plot these values on a scatter diagram and draw a line of best fit.

Estimate the value of y when x is 1·9 and also the value of x when y is 5·0.

4. Here are the marks given to 10 pupils for their work in history:

Pupil	a	b	c	d	e	f	g	h	i	j
Term's work	30	35	35	50	30	75	80	82	85	90
Exam mark	35	32	30	46	35	68	72	70	73	69

Should there be a correlation? Obtain the line of best fit and suggest an examination mark for the pupil who obtained 60 for the term's work.

5. Plot these pairs on a scatter diagram and draw a line of best fit.

(1, 2); (3, 2); (3, 3); (4, 4); (5, 3); (5, 5); (7, 5); (7, 6); (8, 7).

6. A mathematics examination had two papers and each was marked out of 50. Here are the results obtained by 12 pupils:

Name	Paper I (50)	Paper II (50)
Pat	13	12
Janet	18	16
Susan	22	22
Wendy	18	23
Nicola	21	29
Mary	25	31
Judy	28	35
Anita	22	36
Helen	27	40
Tessa	32	44
Elizabeth	31	48
Janette	45	15

Would you expect there to be a direct correlation between the two papers?

Plot the results on a scatter diagram and consider taking special action before attempting to draw a line of best fit.

Inverse correlation

Here are the returns of the number of deck chair tickets sold during a week in August at a seaside resort:

	Mon.	Tues.	Wed.	Thurs.	Fri.	Sat.	Sun.
Number of tickets	45	35	7	14	22	48	32

Here are the rainfall figures at the same place during the same period:

	Mon.	Tues.	Wed.	Thurs.	Fri.	Sat.	Sun.
Millimetres of rain	2·5	5·5	21·5	18	11·5	0	7·5

If we put these together in a table and plot as a scatter diagram we get:

	Mon.	Tues.	Wed.	Thurs.	Fri.	Sat.	Sun.	Mean
Number of tickets	45	35	7	14	22	48	32	29
Millimetres of rain	2·5	5·5	21·5	18	11·5	0	7·5	9·5

This diagram shows a pattern of points which tends to form a line and suggests a correlation as before, but this time the line slopes *downwards* from left to right. This sort of correlation is called INVERSE or NEGATIVE. It implies that an increase in one quantity leads to a decrease in the other, and vice versa.

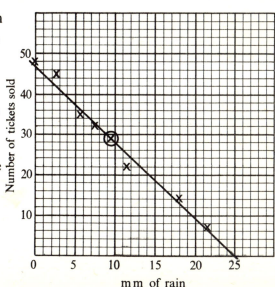

Lack of correlation

A set of results can give a scatter diagram which looks like this:

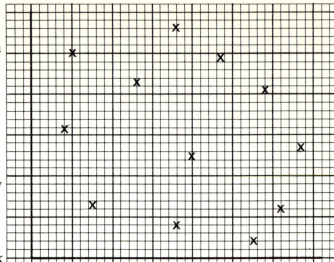

There is no apparent trend or pattern in these points and no correlation exists.

You would not expect there to be any relationship between ability in chemistry and colour of hair.

See if you can think of other examples where correlation is extremely unlikely.

False correlation

Consider this set of imaginary figures:

Year	1967	1968	1969	1970	1971	1972	1973	1974
Sales of chewing gum (£'000s)	1·2	2·1	2·7	3·0	4·5	5·4	6·6	7·1
Number of cases of measles ('000s)	0·8	1·2	2·3	2·8	3·5	5·3	5·9	6·2

If plotted on a scatter diagram, we obtain:

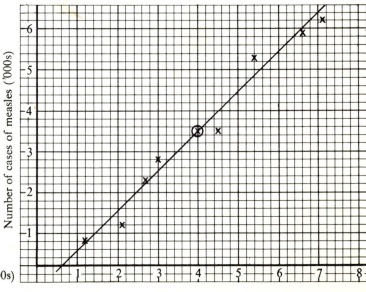

Although a very good line of best fit can be drawn, it would be false to argue that a direct correlation existed between chewing gum and measles. It is true that they have both increased over the years shown, but evidence of their connection must exist before the scatter diagram has any value. It was misleading even to try to draw a scatter diagram of such unconnected data.

A numerical measure

It is possible to calculate a CORRELATION COEFFICIENT between two sets of values. This is arranged so that perfect direct correlation gives +1, no correlation gives 0 and perfect inverse correlation gives −1.

At the beginning we can use this idea in a simple way with our scatter diagrams:

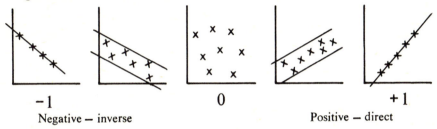

−1		0		+1
Negative — inverse				Positive — direct

Duplicated points

Consider this table of values:

x	0·4	0·9	1·4	2·4	2·4	3·2	3·6	3·9
y	0·7	1·1	2·0	2·7	2·7	3·8	4·0	3·7

You will notice that two entries have the same pair of numbers (2·4, 2·7) and when these details are plotted on a graph only seven distinct points will appear although there are eight pairs of entries in the table.

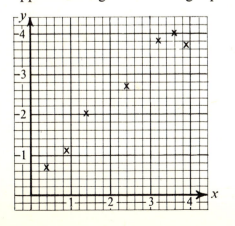

Possible solutions:

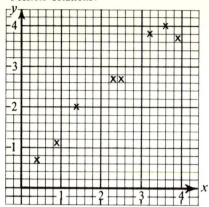
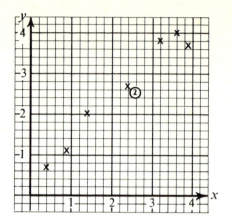

a. Plot two points very close together on either side of the real point.

b. Put a figure 2 in a circle close to the point to indicate that it represents two entries.

The situation could be a lot worse than this and if there were many duplicated points it might not be reasonable to represent the data by a scatter diagram at all.

In certain circumstances a different form of scatter diagram will help. Consider this situation:

	Pat	Mary	John	Dick	Phil	Jane	Jean	Tom	Sandra	Nick
Number of girls in family	2	2	0	1	0	1	1	3	2	2
Number of boys in family	1	1	1	3	1	0	0	2	1	1

This could be represented diagrammatically as:

(Notice that values are placed in the spaces and crosses placed in the squares.)

Such a diagram can at times show a correlation band but it is not possible to draw a line of best fit.

Why is square 0, 0 shaded out?

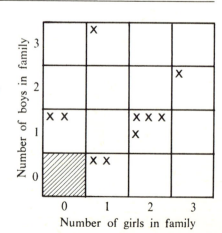

An interesting way of representing duplicated points is to represent all entries by small wooden blocks and then if two come together they can be piled on top of each other and form a three-dimensional scatter diagram.

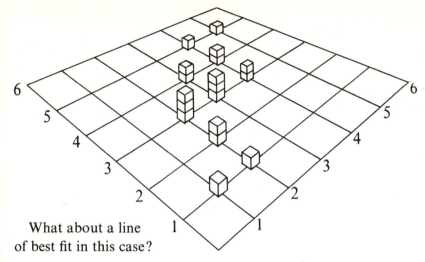

What about a line of best fit in this case?

Exercise B

1. Choose ten pop groups and allocate marks to them, giving 10 marks to the one you think best and so on down to zero if you think there is one which is very bad indeed. Get a partner to do the same thing with the same groups.

Make a table of results and then plot them on a scatter diagram.

See if it is possible to draw a line of best fit.

2. See if you can use the medical-room scales to find the height and weight of every pupil in the form. Plot the results on a scatter diagram and examine it for correlation.

A large diagram can be made on a blackboard or pegboard and instead of plotting points, a small card with each pupil's name on it can be pinned in place.

3. Here are three different sorts of scatter diagrams, and, where possible, the line of best fit has been added:

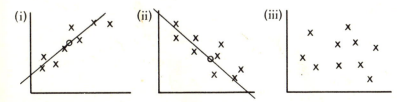

a. Describe each type of correlation shown.

b. Suggest situations which might give rise to each diagram.

4. Describe these two scatter diagrams in words.

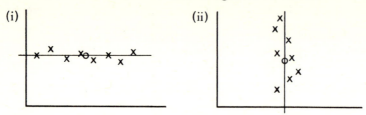

(i)

(ii)

5. Carry out a survey of the number of boys and girls in each of the families of the pupils in your class including yourself. Represent your results by crosses in squares.

6. Represent the results of your survey in question 5 in three dimensions. Use wooden blocks or make cardboard cubes and write the name of each pupil on a block or cube and stick them to the baseboard.

7. What will the scatter diagram look like for this unusual situation?

Pupil	a	b	c	d	e	f
Geography mark	15	37	42	53	60	71
History mark	15	37	42	53	60	71

Try to reach a decision without actually drawing the graph.

8. In the preliminaries for the school sports every boy had to do both track and field events. Here are the results for a group of 10 boys who took part in the discus and the 100 metres.

Boy	a	b	c	d	e	f	g	h	i	j
Discus (m)	31	9	14	14·3	25·3	16·2	20	18·4	20	27·5
100 m (s)	13·5	11·0	11·8	11·9	13·4	12·1	12·5	12·0	12·8	13·9

Plot these results on a scatter diagram and comment on any correlation.

9. Work backwards from the diagram and produce a table of results.

a. What do you think about the correlation of marks as shown by this scatter diagram?

b. If a pupil obtained 65 marks in French but was absent for the geography can you estimate his mark from the graph?

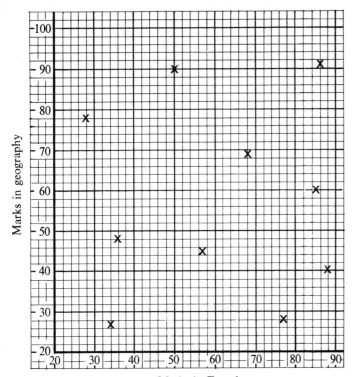

Marks in French

10. If the correlation was calculated for each of the following pairs of quantities, indicate with a plus sign those for which you expect the correlation to be positive, with a minus sign those for which you expect the correlation to be negative (inverse), and with 0 those for which there is no relationship:

a. The amount of rain falling on a certain day and the number of people attending a garden fête on that day.

b. The amount of sweets sold and tooth decay.

c. Height and beauty.

d. Wage rates and the number of cars sold.

e. The circumference of a circle and its radius.

f. The speed of a journey and the time taken to complete the journey.

For which of the examples given above is there a perfect relationship (i.e. a correlation coefficient of +1)?

Give *two* additional pairs of items, for one of which there is positive correlation, and for the other negative correlation, indicating clearly which is positive and which is negative. (AL)

11. A vertical spring, fixed at its upper end, was stretched by the application of weights to its lower end, and the length of the springs for various loads was measured. The results obtained are given in the table on the next page:

Load (g)	0	20	40	60	80	100	120	140	160
Length (cm)	16·0	16·9	18·2	19·0	19·8	21·0	22·1	22·9	24·1

Plot a scatter diagram and draw a line of best fit.

Use your line of best fit to estimate:

a. The load, in grammes when the length of the spring is 16·6 cm.

b. The length of the spring, in centimetres, when the load is 128 g.

Indicate *clearly* on your diagram how you obtained each answer. (MX)

12. Two judges, X and Y, were asked to put eight players in order of merit independently of each other. The player whom they thought was the best was to be marked 1, the next one was to be marked 2, and so on down to 8.

Their assessments were as follows:

Players	A	B	C	D	E	F	G	H
Judge X	1	6	3	7	2	5	8	4
Judge Y	1	5	4	6	2	8	7	3

a. Which players did the judges agree about?

b. Which player did the judges disagree about most strongly?

c. Using 1 cm for each grade in the assessments, draw a scatter diagram for these results.

d. Draw the straight line which best shows the relationship between the assessments (mark this 'Line 1').

e. Draw the lines which would have been obtained:

i. if the judges had agreed exactly about all players (mark this 'Line 2'), and

ii. if the judges had been exactly opposite to each other in their assessment of each player (mark this 'Line 3'). (EA)

4 Frequency distributions

We have met the idea of a frequency distribution already in Chapters 1, 2 and 3. A frequency distribution exists whenever you are counting or measuring and you find more than one item in the same category. For example, here is the result of a survey about the number of lessons spent each week on certain subjects:

Lesson	Number per week (frequency)
English	6
Mathematics	5
Science	5
Geography	2
History	2
Craft	4

From this we would say that English has a frequency of 6 periods per week, mathematics has a frequency of 5 and so on.

In this chapter we shall be concerned with different sorts of frequency distributions.

Rolling dice

Experiment 1

   a. Roll a single die for at least fifteen minutes and make a tally of the scores obtained.

Score	Tally	Total (*frequency*)
1		
2		
3		
4		
5		
6		

Total up and draw a bar chart showing the frequency of each score.

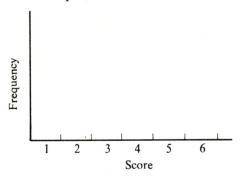

What do you notice about your results?

b. Gather the results from everybody in the class who has done this experiment and represent the grand totals on another bar chart. Is the general shape any different from that obtained in (*a*)? If so, write down your observations.

Experiment 2

a. Work with a partner, one to roll while the other one tallies. Roll two dice each time and note the combined scores.

Carry on for at least fifteen minutes and then total and draw a bar chart as before.

Do you notice anything about your chart this time? How does it differ in shape from that obtained in Experiment 1? Does any score occur more frequently than the others?

b. All the class should pool results and everybody should draw a chart of these combined totals. Make a note of observations along the lines suggested in (*a*).

Experiment 3

Use three dice this time. Work with a partner as before.

a. Roll, record and draw a chart of your own results. List any observations.

b. Pool all available results, draw a chart and record observations.

The combined results can be described by the following frequency diagrams:

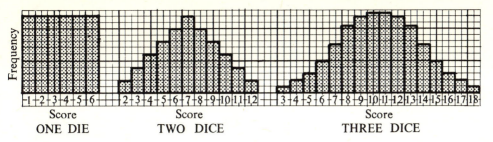

Score	Score	Score
ONE DIE	TWO DICE	THREE DICE

Or, to make the representation even easier:

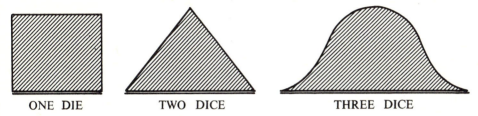

ONE DIE	TWO DICE	THREE DICE

We call the one-die shape a RECTANGULAR DISTRIBUTION, the two-dice shape a TRIANGULAR DISTRIBUTION and the three-dice shape a BELL-SHAPED DISTRIBUTION.

What explanation could you give if the result of rolling one die gave this general shape?

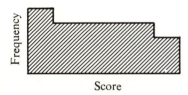

Score

How do you think you could produce an actual set of results which would give this frequency diagram?

Experiment 4

See if you can obtain a wooden cube with length of edge 3 cm or more. Put numbers 1 to 6 on its faces and into one face hammer a number of very small nails such as tin-tacks or gimp pins.

Roll for at least fifteen minutes and examine your results graphically as before.

State any conclusions you reach.

An alternative way of making a loaded die is to take an ordinary die and drill out the spots on the six-face, going nearly through to the other side. Refill three of these in a row with strips of lead or solder.

Spinning tops

Experiment 5
Construct this regular octagon on thick card, or better still, plywood. Start with a square of side 6 cm.

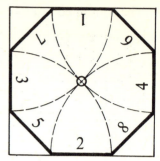

Put the numbers 1 to 8 on the sides, drill a hole through the centre and insert a piece of sharpened dowel, or a pencil about 6 cm long. Put some glue at the joint and work the dowel about so that some of the glue gets into the hole. Leave to dry.

Carry out a statistical investigation to find out how accurate your construction of this top has been.

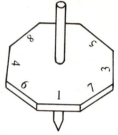

Experiment 6
Make a hexagonal top in the same way as the octagonal one in Experiment 5. Start with a circle of radius 4 cm. Put lots of extra glue along one edge to weight it. Draw a frequency diagram of the results of very many spins.

Comment on your results and describe the shape of your chart.

Exercise A

1. The sizes of shoes in a group of boys were recorded as:

Size of shoe	6	$6\frac{1}{2}$	7	$7\frac{1}{2}$	8	$8\frac{1}{2}$	9	$9\frac{1}{2}$	10	$10\frac{1}{2}$
Number of boys	1	3	6	10	12	11	9	5	2	1

Represent these results by a bar chart and try to fit a smooth curve which gives the general shape of the distribution.

2. The GCE results in English for a school during a certain year were as follows:

Grade	1	2	3	4	5	6	7	8	9
Number of pupils	1	1	2	4	8	14	16	12	7

Draw a bar chart of these results and on it draw a smooth curve which gives the general shape of the distribution.

3. Here are some more GCE results, but this time the number of subjects passed is given:

Number of subjects passed	1	2	3	4	5	6	7	8	9	10
Number of pupils	9	18	14	10	8	5	2	1	1	1

Draw a bar chart of these results and on it draw a smooth curve which gives the general shape of the distribution.

4. Here are the number of letters delivered one morning by a postman in a certain road:

House numbers	1	2	3	4	5	6	7	8	9	10	11	12	13	14	15	16
Number of letters delivered	2	0	3	1	7	1	0	0	0	3	0	2	1	1	1	4

Represent this information on a bar chart. Is it possible to fit a curve of general shape to this distribution?

Skewed distributions

If you have answered Exercise A, question 1 correctly, you should have:

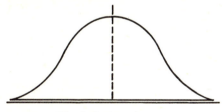

The bell-shaped curve

However, questions 2 and 3 will give:

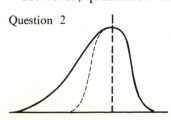
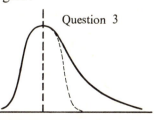

These last two are very similar to the one above except that their ends have been pulled out on one side. Such cases are called SKEWED DISTRIBUTIONS.

The left tail of Number 2 has been pulled out; it is said to be skewed to the left.

Number 3 is skewed to the right.

These distributions are sometimes said to be negatively skewed (2) and positively skewed (3). Can you see why?

Bombing

Take a large sheet of paper and rule up a target area as follows; use plain coloured paper so that it is easier to see the rice. Draw the parallel lines 2 cm apart.

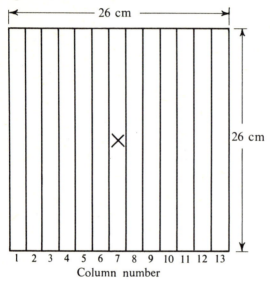

Count out exactly 100 grains of rice, take a few in the fingers each time and gently bomb the target in the middle. When all the rice has been used up, record the totals falling in each column. Repeat six times and draw a graph of the combined results.

What shall we do with those which lie on a line? There are three possible decisions:

a. Count $\frac{1}{2}$ for each column.

b. Decide on one column or the other according to the size of the overlap.

c. Decide before you start to put them in either the lower or the higher column.

Repeat the experiment from different heights. Heights of 15 cm, 12 cm and 10 cm are suggested.

What about the grains which bounce right off the target area?

The results should be bell-shaped distributions.

See if you can now obtain a skewed distribution. Possible ways are:

a. Incline the target board slightly by raising either the left or righthand edge.

b. If you usually drop the rice with your left hand try dropping it with the right or vice versa.

c. Let your hand travel slowly across the board from left to right or vice versa, while you try to release your bombs over the centre of the target.

Other distributions

Consider the results of this survey:

Number of times been abroad for holiday	0	1	2	3	4	5	6	7
Number of pupils	20	10	4	3	2	2	1	1

When this is represented on a bar chart we get:

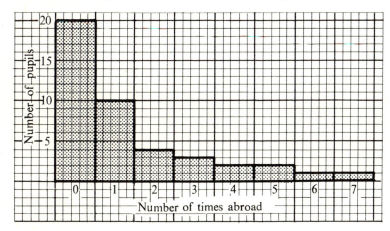

This can be represented by:

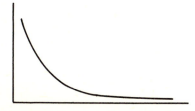

There are many other distribution shapes as well as those we have already encountered. Some more will be met in the next exercise.

Bell-shaped distributions

The bell-shaped distribution is an extremely common one. It is frequently produced as a result of a survey concerning estimating, weighing or measuring and because of this it is usually called the NORMAL DISTRIBUTION CURVE. A great deal of use is made of this shape in later work.

Experiments which might produce normal distributions

1. Measure the heights (in centimetres) of a large number of pupils (say 100) all of about the same age. Measure to the nearest cm and present the results as a bar chart. Try to fit a curve giving the general shape.

2. As in question 1, but this time obtain the weights, to the nearest kilogram.

If questions 1 and 2 are carried out with the same pupils, a scatter diagram and line of best fit can also be drawn.

3. If you ever have a competition to guess the weight of a cake at the school fête, the guesses might be worth examining for a normal distribution.

4. Take a ball of string, measure and cut off a piece 10 cm long. Throw this piece away and then continue cutting off pieces which you estimate are 10 cm long, until you have a large number of pieces. Measure them all (to the nearest mm) and present your results as a bar chart. Only one person should do the cutting, but other members of the class can be measuring and recording the results as they come. Try combining the results of two pupils and see if there is any difference in the general shape.

5. Obtain the examination results for as many pupils as possible in the same examination. Group the marks into blocks of ten marks, i.e. 1–10, 11–20, 21–30 and so on, assuming that nobody got 0.

Present as a bar chart. Fit a general shape as before and comment upon it.

Exercise B

1. Match these tables of frequencies with the general shapes given below:

x	*1*	*2*	*3*	*4*	*5*	*6*	*7*	*8*	*9*	*10*	*11*	*12*
a	20	10	3	2	1	1						
b	50	60	48	30	12	6	2	1				
c	45	30	18	9	3	8	20	29	37			
d	1	2	4	7	12	14	11	6	3	1		
e	1	2	3	4	5	4	3	2	1			
f	8	7	8	8	9	7	8					
g	1	2	4	7	10	18	20	18	16			
h	2	5	6	11	20	50						
i	1	3	8	12	7	3	3	6	11	9	4	2

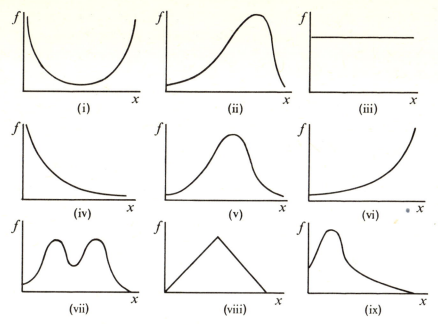

(i)	(ii)	(iii)
(iv)	(v)	(vi)
(vii)	(viii)	(ix)

2. Make a model of a bell-shaped distribution curve by cutting lengths of balsa wood or card strips to fit this outline:

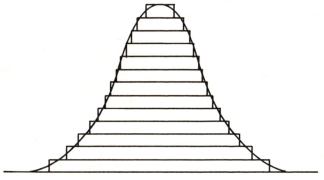

The thinner the strips, the more accurate the model will be.

Use and display your model to illustrate distributions which are:

(a) Negatively
   skewed

(b) Positively
   skewed

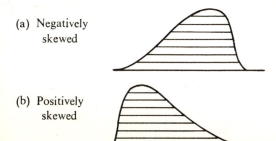

3. Can you think of a set of frequencies, or better still, an actual situation, which would have a distribution corresponding to this general shape?

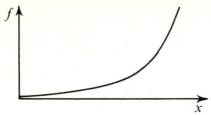

4. The bell-shaped distribution curve has LINE, or BILATERAL SYMMETRY. This means that you can fold it along the dotted line and one half will exactly match the other.

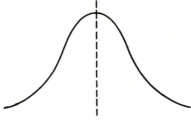

a. Use this fact to help you draw such a curve.

b. Which other curves in question 1 have this property?

5a. What shape of frequency distribution would you expect from a survey conducted to find the number of children per family in 100 families?

b. Carry out such a survey in your own school and compare your result with your answer to (*a*).

6. A survey of the number of children per family in a large number of families gave this shape of distribution:

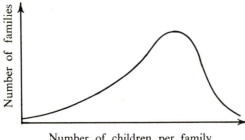

Number of children per family

Suggest possible countries where this survey might have been carried out.

7. Carry out a survey in your own class on the time it takes to get to school in the morning. Work to the nearest minute.

a. Group people in blocks of five minutes:

1–5; 6–10; 11–15; 16–20 and so on.

b. Group people in blocks of ten minutes:

    1–10; 11–20; 21–30 and so on.

Represent each set of results on a separate chart and comment on the shapes.

8. If a survey was carried out to find the number of pupils having school dinner each day for a week, what shape of distribution would you expect? Check your conclusions with the figures for your own school.

9. Select a page from your local telephone directory and tally how many numbers end in 0, 1, 2 and so on up to 9. What sort of distribution shape do you expect? Do you actually obtain this?

10. Carry out the same sort of investigation as in question 9 but this time count the final digit in car registration numbers.

The quincunx

Here are the plans for a piece of apparatus which produces the shape of the normal distribution curve:

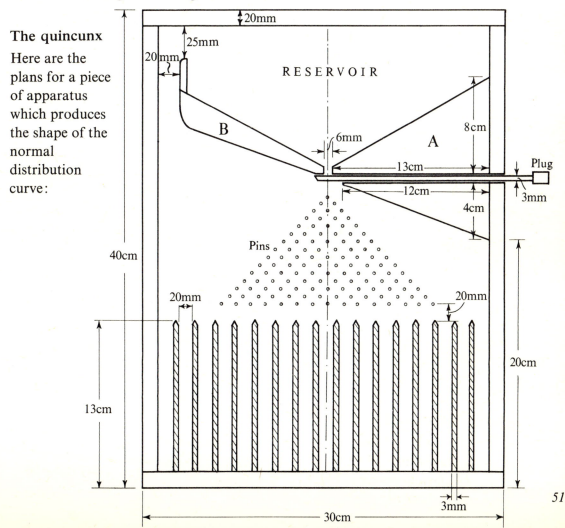

Full-size detail of pin layout:

Mark the positions of the nails on a piece of paper.

Place the paper in position on the base-board and nail through it.

Tear off the paper when finished.

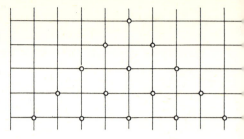

MATERIALS

1 piece 9 mm plywood, 40 cm × 30 cm for baseboard.

1 piece of 3 mm perspex, 40 cm × 30 cm for front cover.

2 m of 20 mm × 40 mm strip wood for border.

16 pieces, each 13 cm long, of 3 mm × 10 mm strip for compartments.

1 piece 18 cm long, of 3 mm × 10 mm for plug.

1 piece of 9 mm ply about 25 cm × 8 cm to make the reservoir pieces.

20 mm panel pins with small heads.

30 g rape seed.

ASSEMBLY

Nail and glue border to baseboard. Mark out all positions on baseboard and be sure to put a centre line right down it.

Cut compartment sides to length and sharpen the ends. Glue edgewise to baseboard.

Drive in nails as detailed above.

Cut two pieces to the shape of A, place on the board in positions A and B and hence obtain the shape of B. When assembling the reservoir make sure that the pieces on the right pinch the plug strip and so keep it tight.

Have a few trial runs to get the correct amount of seed so that the central column is just full and a good normal curve is obtained.

Screw on the front.

5 Running totals

There are two ways of carrying out an addition:

1. When given three numbers to add together such as

21 + 16 + 18,

we can write them out in columns of tens and units:

```
    t   u
    2   1
    1   6
+   1   8
   ___ ___
    5   5
```

and add up the units. (1 + 6 + 8 which comes to 15). Write the 5 in the units column and carry the ten giving 5 in the tens column. Answer, 55.

This is the usual way to add in the head with the help of paper and pencil.

2. We can represent the addition of 21 + 16 + 18 as steps along a number line:

The answer of 55 can be read off from the line: notice that the intermediate totals of 21 and 37 can be seen as well.

This second method represents the way the addition would be done with a desk calculator.

53

Adding by desk calculator

To do the sum $21+16+18$ by a desk calculator you would follow this procedure:

a. Enter 21 into the setting register (SR); turn the handle and 21 goes into the accumulator (A).

b. Enter 16 into SR; turn the handle and 16 is *added* into A which now reads 37.

c. Enter 18 into SR; turn the handle and 18 is *added* into A which now reads 55.

The accumulator makes a running total, number by number, and it tells you how the sum is growing as each number is added on.

Here is another example: Add $252+629+81+103$.

	Entries in SR	Reading of A
a.	252	252
b.	629	881
c.	81	962
d.	103	1065

The flow diagram for this process is very elementary:

This is the way a computer adds a lot of numbers together; it deals with them one by one.

Book-keepers and bank accountants use running totals so that the amount of money in hand at any one time can be immediately seen.

Exercise A

1. Copy and complete this table:

Number	Running totals
127	
31	
438	
226	
407	

Now add up the numbers 127, 31, 438, 226, 407 in the usual way by units, tens and hundreds. Can you see a way of checking your running total?

2. Put these numbers into a table as in question 1 and make running totals. Check your results by direct calculation:

$$11\cdot6, \ 4\cdot7, \ 18\cdot2, \ 9\cdot48, \ 15\cdot5, \ 10\cdot09.$$

3. During a week's holiday a boy worked at fruit picking and his earnings,
day by day, were:

Monday	£1·25
Tuesday	£1·35
Wednesday	£1·75
Thursday	£1·60
Friday	£0·75

Make running totals of these earnings.

4. The stages on the London to Brighton road are:

London–Croydon	16 km
Croydon–Redhill	18 km
Redhill–Crawley	16 km
Crawley–Brighton	37 km

Make a table of running totals and hence show the distances:

London–Croydon
London–Redhill
London–Crawley
London–Brighton

5. Here are the number of pupils staying to school lunch each day during
a week:

Monday	186
Tuesday	194
Wednesday	190
Thursday	182
Friday	201

a. Make running totals and check by direct addition.

b. How many meals had been served up to and including Wednesday?

6. See if you can obtain from the school office the school-dinner totals for
each day during a week and make your own running totals.

Can one running total be less than any running total before it?

7. The sale of tickets for a school play went like this:

	Monday	47
	Tuesday	169
Week 1	Wednesday	203
	Thursday	292
	Friday	121
Week 2	Monday	29
	Tuesday	7

When had a quarter of the tickets been sold?
When had half the tickets been sold?

Making graphs of running totals

Consider the situation in which the number of pupils late for school during one week was:

Monday	Tuesday	Wednesday	Thursday	Friday
2	4	6	5	3

We can represent this on a bar chart in the usual way:

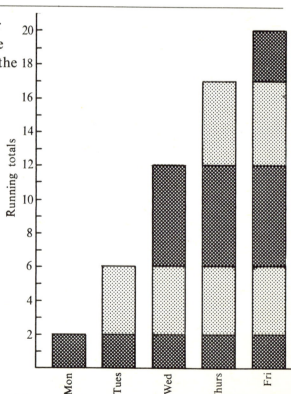

The running total table will be:

Day	Number late	Running totals
Monday	2	2
Tuesday	4	6
Wednesday	6	12
Thursday	5	17
Friday	3	20

If we pile the bars of the bar chart on top of one another we shall also be able to represent the running totals graphically:

The general shape of this result is like a stretched out S.

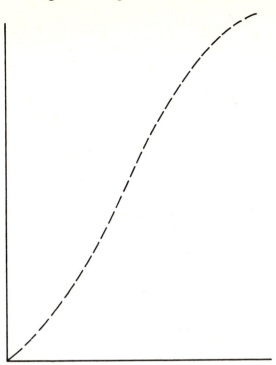

Example

Here are the marks obtained by a class of pupils in a test:

Mark	1	2	3	4	5	6	7	8	9	10
Number of pupils obtaining that mark	1	2	2	5	6	8	5	3	1	1

Make a table showing the running totals.

Note carefully the way in which each group has been described; for example, the group obtaining 5 marks or fewer includes all the pupils who have scored 1, 2, 3, 4 or 5 marks. Another way to describe this group would be 'up to and including 5', or even, 'fewer than 6'.

Marks	Running total Number of pupils
Only 1 mark	1
2 marks or fewer	3
3 ,, ,, ,,	5
4 ,, ,, ,,	10
5 ,, ,, ,,	16
6 ,, ,, ,,	24
7 ,, ,, ,,	29
8 ,, ,, ,,	32
9 ,, ,, ,,	33
10 ,, ,, ,,	34

Plot the graph:

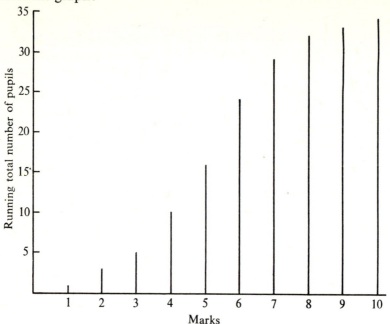

Again notice the general shape. It is very much like that in the previous example.

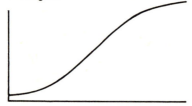

There is a special name for this stretched S, it is called an OGIVE (pronounced 'oh-jive'). A curve of this general shape is always obtained when a running total is plotted on a graph. Sometimes it will be more stretched, compressed, flattened, curved or even stepped, but it will still be the same general shape.

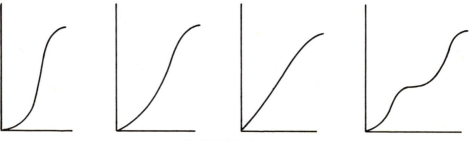

O G I V E S

All continue upwards; they never go down.

Exercise B

1. See if you can obtain the figures giving the number of pupils late (or absent) each day at school for a week. Represent this information first as an ordinary bar chart showing the total for each day, and second as a running total diagram by piling the bars on top of one another as in the text example.

2. Here are the totals in a class having school milk each day for a week.

Monday	*Tuesday*	*Wednesday*	*Thursday*	*Friday*
12	14	17	16	10

Make a running-total type of bar chart to display this information.

3. Make a table showing the running totals of these marks and plot your results on a graph by putting a cross to show each total. Sketch the general shape obtained. Is it an ogive?

Marks	0	1	2	3	4	5	6	7	8	9	10
Number of pupils (frequency)	1	1	2	4	8	8	11	6	3	2	1

4. What sort of running-total ogive will this set of results give? See if you can decide on the shape before you actually draw the graph.

Score	1	2	3	4	5	6
Number of times (frequency)	3	3	3	3	3	3

5. Make tables, running totals, graphs and general shapes for each of the following parts. Part (*a*) has been done for you as an example.

a.

x	1	2	3	4	5	6	7	8
f	1	2	4	8	11	19	20	16

d.

x	1	2	3	4	5	6
f	20	11	3	2	1	1

b.

	1	2	3	4	5	6	7	8	9
f	1	2	4	8	13	10	7	3	1

e.

x	1	2	3	4	5	6	7	8
f	25	10	4	3	2	5	9	22

c.

x	1	2	3	4	5	6
f	1	4	6	12	20	45

Will all the parts (a) to (e) give general shapes which can still be called ogives? Which gives the most typical S shape?

The answer to 5(a):

x Score	f frequency	Running totals
1	1	1
2	2	3
3	4	7
4	8	15
5	11	26
6	19	45
7	20	65
8	16	81

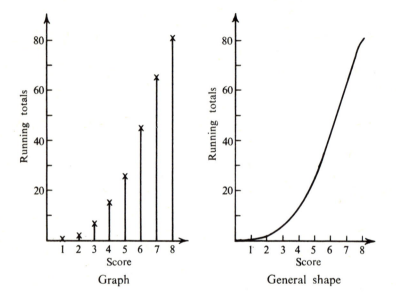

Graph

General shape

6. When playing darts you start by writing 101 on the score board and subtracting your score from this each time until zero is reached.

Here is part of a darts score board:

How many did John score each time?

Draw a graph to show how John made his total score of 101.

6 Probability

Why are both captains prepared to accept the ruling of a coin when they toss at the beginning of a match?

They know that the coin must land either head up or tails up and that any such result will be quite fair to both sides. If the coin did land on its edge in the grass, both would agree to a second toss: it is extremely unlikely that the same thing would happen again. Neither captain on the other hand would be happy to toss with a bent coin or one with two heads.

Neither captain would be prepared to say that the coin must fall head up just because last time they tossed it fell tails up: neither captain is able to predict with absolute certainty the result of a single toss, and yet each will agree that, in the long run, the number of head outcomes will be about the same as the number of tail outcomes.

We are now going to do some experiments to see whether the captains are right in their acceptance of these ideas. Some new words will be used and each will have its own particular purpose in explaining the captains' assumptions.

Experiment 1

Everyone in the class tosses a coin 10 times and records each result. This is called the outcome of the individual experiment and can be recorded as, for example, 6 heads and 4 tails.

Display these individual outcomes as a group outcome in a table like the one shown overleaf:

Name	Heads	Tails
G. Cox	4	6
P. Warner	5	5
⋮	⋮	⋮
E. Williams	6	4
P. Palmer	7	3
Totals	147	153

The table below has been made up for a supposed class of 30 pupils.

When these outcomes are tallied, the result is:

The outcome of your class experiment will not agree exactly with this result, but it should now be clear that although the frequency of the 5–5 outcome will be large, it need not be the most frequent. Nevertheless, the total outcome of the class experiment, 147–153, does explain the belief of the two captains.

Heads	Tails	Tally	Total
10	0		0
9	1	I I	2
8	2		0
7	3	I	1
6	4	⊦⊦⊦ I I I	8
5	5	⊦⊦⊦ I I I	8
4	6	⊦⊦⊦	5
3	7	I I I	3
2	8	I I	2
1	9	I	1
0	10		0
			30

In the long run, the number of heads and tails will be approximately equal, in spite of the fact that individual pupils may have outcomes that show a considerable difference from the expected equal one of 5–5.

The results of this experiment are very important. In any experiment there are two possible outcomes—the outcome that is expected, and the outcome that actually happens, and these may not be exactly the same.

The outcome that is expected and the outcome that occurs may not be exactly the same

62

In the case of the coins, we expect equal numbers of heads and tails; the numbers we actually get will not be quite equal, although the more times we repeat the experiment, the closer the two outcomes will be.

To describe the expected outcome of an experiment, we use what is called the PROBABILITY of the outcome. In the case of G. Cox, for example, the probability of his outcome is 5 in 10 to 5 in 10, that is $\frac{5}{10}$ to $\frac{5}{10}$ or $\frac{1}{2}$ to $\frac{1}{2}$.

His actual outcome in this case is $\frac{6}{10}$ to $\frac{4}{10}$.

The expected probability of the class experiment as a whole is $\frac{150}{300}$ to $\frac{150}{300}$ or, once again, $\frac{1}{2}$ to $\frac{1}{2}$ although the actual outcome is $\frac{147}{300}$ to $\frac{153}{300}$.

If the whole experiment was repeated, the expected probabilities would not change, but the actual outcomes might well be a little different, but we would be surprised if the whole result was *very* different from the expected outcome of $\frac{1}{2}$ to $\frac{1}{2}$.

Notice that the probabilities are expressed as fractions; they are usually reduced to their lowest terms for convenience, and very often only one of the two probabilities is mentioned. Thus, for example, we would talk about the expected probability of a head being $\frac{1}{2}$, implying at the same time that the probability of a tail is $\frac{1}{2}$.

We also talk about an EXPERIMENTAL PROBABILITY, which is the result, expressed as a fraction, from the combined results of a large number of experiments. An experimental probability should always agree fairly closely with the expected probability, provided enough experiments have been done.

If the two values do not agree, it is often an indication that something may be wrong with the experiment, for example the coin may not be true.

Experiment 2

What is the expected probability of throwing a 5 with a die?

What is the expected probability of throwing an even number with a die?

Throw a die until you get a 5. Throw a die until you get an even number. What are your actual outcomes in each case?

The whole class should try both of these and pool their results to give the experimental probability for each case.

Do the experimental results agree with the expected results closely enough to justify the values of the expected probabilities?

Experiment 3

There are 52 playing cards in a pack. Work out the expected probabilities for the following outcomes:

a. Drawing the 7 of diamonds.   d. Drawing a card less than 5,
b. Drawing any heart.   if an ace counts as high.
c. Drawing an ace.   e. Drawing a black card.

Shuffle a pack of cards, and draw the top card; return this to the pack and reshuffle if it is not the 7 of diamonds. Repeat until the 7 of diamonds is drawn.

What is *your* actual outcome for the experiment of drawing the 7 of diamonds?

Find your actual outcomes for each of the above experiments.

Determine the experimental probability for the whole class in each case. Do the results justify the assumptions you made in calculating the expected probabilities?

Experiment 4

What is the probability that a person you meet will have a birthday on the same day of the year as you?

What is the probability that a person you meet will have a birthday in the same month as you?

Find out within your class how the expected probabilities compare with the actual probabilities.

Experiment 5

If any letter is chosen from the word 'leopard', what is the expected probability that it will be a vowel?

Write these seven letters on equal pieces of card, shuffle the cards, and obtain some actual outcomes.

Class discussion

Two cricket captains use a coin to decide who should have the choice of batting. What other methods could be used and still give a fair decision in the long run? Make a list of all the ways. Allocate each way to a group of pupils and carry out experiments to determine the actual fairness of each method.

Probability of 0 and 1

Phrases like 'not in a month of Sundays', 'a dead certainty', 'not a ghost of a chance' and many similar expressions, are all trying to convey ideas which are very simple to describe in mathematics.

What is the probability of getting a tail with a double-headed coin? The probability outcome in this case is 0 for a tail and 1 for a head. Any outcome which must happen is said to have a probability of 1; an outcome which cannot possibily happen is said to have a probability of 0.

Any outcome that is not very likely to happen, like winning the treble chance on the football pools, has an expected probability that is so small that

we often take it to be 0; an outcome that is almost certain to happen, like getting run over if we cross a busy street blindfolded, has an expected probability of almost 1, and we usually take this as 1.

Class discussion

1. Assign probabilities to the following events:
Pigs will fly. The sun will rise tomorrow. It will snow on 1 August. Tomorrow is Thursday. A score of 1 with a pair of dice.

2. A probability scale. Make a large copy of a scale like this and display it on the wall.
Why will every event give rise to two points on the scale? Add as many extra points as you can.

1 — Certainty

$\frac{3}{4}$ — Not drawing a diamond from a pack of cards

$\frac{1}{2}$ — Coin will be tails

$\frac{1}{4}$ — Drawing a diamond from a pack of cards

$\frac{1}{13}$ — Drawing an ace from a pack of cards

0 — Impossibility

Combination of events

Many events in real life are combinations of one or more separate happenings. The probability that I will not be late for school on a particular morning is affected by many other probabilities, such as the probability that I wake up in time, the probability that the toast is not burnt, the probability that my bus is on time or that my bicycle does not have a flat tyre, the probability that the school clock is right, and so on.

It is not easy to see exactly how these individual probabilities affect the final probability of being late: the train being late, or my over-sleeping badly, will make a great difference, while breaking a shoelace will have only a slight effect.

We will now study how the probabilities of individual events can affect the probability of a combined result.

Example

A coin and a die are tossed together.

    a. What is the probability of scoring a head and a 5?

    b. What is the probability of scoring a tail and a 3?

    c. What is the probability of scoring a tail and a number greater than 2?

It is helpful in such questions to arrange all possible outcomes in a table.

Outcome of coin	*Head*	H:1	H:2	H:3	H:4	H:5	H:6
	Tail	T:1	T:2	T:3	T:4	T:5	T:6
		1	2	3	4	5	6

Outcome of die

Each box represents a possible outcome; T:3 stands for the outcome of a TAIL and a THREE, H:5 stands for the outcome of a HEAD and a FIVE, and so on.

We can see that there are twelve possible outcomes; this is not surprising since there are two possible outcomes for the coin and six possible outcomes for the die.

    a. The outcome of scoring a head and a 5 is *one* of these twelve possible outcomes; the probability of scoring a head and a 5 is $\frac{1}{12}$.

    b. The outcome of scoring a tail and a 3 is *one* of these twelve possible outcomes; the probability of scoring a tail and a 3 is $\frac{1}{12}$.

    c. The outcome of scoring a tail and a number greater than 2 is a combination of the four outcomes T:3, T:4, T:5 and T:6. The probability of scoring a tail and a number greater than 2 is $\frac{4}{12}$ or $\frac{1}{3}$.

Another problem of great importance is the calculation of the final probability of an outcome which consists of several distinct stages.

Example

I am allowed to throw a die twice.

   a. What is the probability that I throw a 6 followed by a 1?

   b. What is the probability that my total score is 7?

	1	2	3	4	5	6
6	(1, 6)	(2, 6)	(3, 6)	(4, 6)	(5, 6)	(6, 6)
5	(1, 5)	(2, 5)	(3, 5)	(4, 5)	(5, 5)	(6, 5)
4	(1, 4)	(2, 4)	(3, 4)	(4, 4)	(5, 4)	(6, 4)
3	(1, 3)	(2, 3)	(3, 3)	(4, 3)	(5, 3)	(6, 3)
2	(1, 2)	(2, 2)	(3, 2)	(4, 2)	(5, 2)	(6, 2)
1	(1, 1)	(2, 1)	(3, 1)	(4, 1)	(5, 1)	(6, 1)

Score on second throw (vertical axis)   *Score on first throw* (horizontal axis)

The above table shows that there are thirty-six possible outcomes altogether.

   a. Throwing a 6 and then a 1 is *one* of these outcomes. The probability is thus $\frac{1}{36}$.

   b. There are six ways in which the total score is 7:

$$\{(6, 1), (5, 2), (4, 3), (3, 4), (2, 5), (1, 6)\}.$$

The probability of scoring exactly seven is thus $\frac{6}{36}$ or $\frac{1}{6}$.

Bookmakers' odds

When a horse is quoted as having odds of 5 to 1, this is a measure of the probability of that horse winning, but with a slightly different twist. If you back the horse with £1, the bookmaker is at the same time prepared to put down £5; the total sum involved is £6 and the winner takes the money.

   You have a $\frac{1}{6}$ stake, the bookmaker has a $\frac{5}{6}$ stake, in the total wager.

   Although the horse is quoted at 5 to 1, your probability of winning is $\frac{1}{6}$, the bookmaker's probability is $\frac{5}{6}$.

   Odds of 5–1 imply that you have one chance of winning and five chances of losing: a probability of $\frac{1}{6}$ implies there is one favourable outcome (you winning) out of a total of six possible outcomes.

   A horse quoted as 'evens' means odds of 1–1; if you put £1 down, the bookmaker covers it with £1, and the winner takes £2. The probability of winning is $\frac{1}{2}$ and is even with the probability of losing, which is also $\frac{1}{2}$.

Exercises

1. Here are some starting prices. Give the probability of winning in each case:

Updown Park	4·30	Fillies' Stakes
Lonely Lady		5–1
Just Janie		10–1
Pretty Polly		100–8
Kay's Delight		evens
On The Beach		33–1
Veronica		2–1 on

(2–1 on means that for every £2 you stake, the bookmaker stakes only £1.)

2. Draw up a table to show the possible outcomes of tossing a red and a blue die together.

What is the possibility of a combined score of 2, of a combined score of 7, of a combined score of 11, with these dice?

3. Draw up a table to show the possible outcomes of tossing a penny and a fivepenny piece together.

What is the probability of both coins showing a head?

What is the probability of the penny showing a head and the fivepenny showing a tail?

What is the probability of the coins showing one head and one tail?

4. See if you can invent a table to show the possible outcomes of tossing a penny, a fivepenny and a tenpenny together.

How many possible outcomes are there?

What is the probability that all three coins will show a tail?

5a. I throw a die. What is the probability that I shall throw (*i*) a 5; (*ii*) at least a 5; (*iii*) a 5, followed by a 4?

b. Given a well-shuffled pack of playing cards, what is the probability that if I select one card it will be (*i*) a spade; (*ii*) an honour card? (Ace, King, Queen, Jack are honour cards.)

6. Mark the faces of an hexagonal pencil with the digits 1, 2, 3, 4, 5, 6.

Carry out an experiment to determine the probability of any particular score.

Could this pencil fairly replace a die in any game requiring one?

7. An old ballad tells of the decision of a shipwrecked crew to eat one of their number. The volunteer was selected by the method of drawing the short straw. Find out what this method is, divide the class into crews of about five each, and, by using matchsticks, determine whether this method is fair.

8. Tossing a coin a large number of times is a long and noisy process. A Probability simple mass-production device which achieves the same result is simply to throw a handful of, say 10, pennies into a cardboard provisions box placed on its side. Throw the coins so that they hit the back of the box and fall inside; the lower flap can be raised to prevent coins rolling out and can be laid flat for sorting.

Verify that such a method does give the same overall pattern as tossing 10 coins separately.

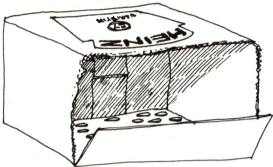

9. *Octahedral dice.* The regular octahedron has eight faces and each one of them is an equilateral triangle.

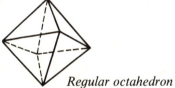

Regular octahedron

a. Copy this net onto thin card, cut out, crease and stick the flaps in to make a model.

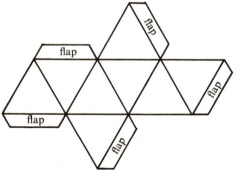

b. Copy the net overleaf onto cartridge paper, cut out and crease all the lines. This will then plait to form a regular octahedron: start your plait with *a* and *b*.

Finish by tucking the last triangle in to make a rigid model. 69

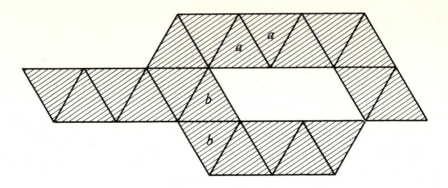

c. Mark the faces of each of your models with the digits 1, 2, 3, 4, 5, 6, 7, 8. Will these be fair eight-sided dice?

Test your decisions experimentally.

10. The final experiment. A revolver has six chambers; Russian Roulette is played by putting a cartridge in one chamber, spinning the chambers, pointing the gun at your head and squeezing the trigger.

If the loaded chamber comes to rest in line with a barrel, the gun fires. What is the expected probability of a success?

Successful for whom?

This should not be carried out as a class experiment!

7 Sampling

Bears like honey, and Pooh is no exception. In the picture we can see Pooh taking his big decision of the morning—which honey pot will be today's lunch? He has discovered, although a 'Bear of Very Little Brain', a very common technique known as *sampling*. His survey of the quality of the honey is carried out by dipping into each pot, and licking his paw. He has realised that he does not have to eat the complete contents of each pot in order to judge its quality. The fact that the quality of a small part can indicate the quality of the whole is fundamental to the idea behind taking a sample.

Many examples of sampling occur: cheese and wine tasters use scoops to take a small sample from the centre of the cheese or from the wine barrel; a factory making electric light bulbs will only test to destruction a few of its daily output; our traffic survey lasted only about twenty minutes and not the whole day; we did not plan to ask every householder about his type of residence, but only the sample that we met outside the Town Hall.

There are two main reasons for examining only a sample rather than the whole:

1. There is the fact that in examining the whole population, as it is called, we may destroy all of what we are examining. This happens in the cases of Pooh's honey, the cheese and wine tasting, and the electric light bulbs. There is no sense in deciding the quality of something if there is nothing left afterwards.

2. The second reason is the occasional difficulty of reaching the whole population—it would be impossible to survey traffic all day and night without a great deal of effort; to visit each householder would not be impossible, but it would be a very long process.

Sampling within an overall population is an extremely important idea in statistics, so we will now do some experiments to find out how reliable the method is, and what precautions we ought to take in order to get a trust-worthy result from a sample.

In this investigation we will use coloured beans, and take samples of about 100 beans from a total population of about 1000 beans, but we must first prepare the experiment:

Experiment 1. The bean experiment
Apparatus required for each working group:
1 plastic or enamel bowl (any round container will do; a small mixing basin is ideal).
1 matchbox tray.
250 g packet of dried haricot beans.
Red and blue ink.
A small container in which to dye the beans—a jam-jar will do.

Preparation. (This could be done at home if necessary.) Count out exactly 100 beans—a level matchbox usually contains almost exactly 100 beans, but the number will have to be checked.

Dye these by dropping them into a small quantity of red ink in the jam-jar. It is not necessary to immerse them in a lot of ink, just a few drops are sufficient if you shake the beans for a minute or so. Do not leave them too long or they might soften and swell. Remove and spread out on newspaper to dry over-night.

Similarly dye exactly 200 beans in blue ink.

Other, and more permanent dyes are shoe dyes, hair dyes and spirit-based wood dyes. Felt-tip pen ink can be made to flow a little and the whole pen used to stir the beans up. In this way you can get other colours than just red and blue and this makes it possible to devise more interesting experiments. We, however, will restrict ourselves to red, blue and uncoloured, or white, beans.

Method. Place all the beans, the 100 red, the 200 blue and the remaining white, in the basin, and mix them by hand.

Using the matchbox tray as a scoop, take a sample of about 100 beans; spread these out on the desk, and sort them into the three different colours.

Assuming that there are very nearly 1000 beans altogether, (you can count them if you like, but it takes a long time), how many red, white and blue beans do you expect to find in a sample of 100? Do you expect to find exactly the numbers in each sample that you take?

If you take ten such samples, do you think the average number of red beans in each sample will be more reliable than the number in any one sample? Will the same be true for the blue and white beans as well?

Repeat the experiment ten times, returning each sample to the basin after you have counted the exact number of red, blue and white beans, and thoroughly stir the whole population each time.

Record your results in a table like this:

Sample number	Red	Blue	White	Total
1	13	18	67	98
2	12	24	70	106
3	16	28	66	110
4	5	18	63	86
5	12	20	71	103
6	10	27	66	103
7	8	22	77	107
8	7	21	74	102
9	7	24	67	98
10	6	18	67	91
Totals	96	220	688	1004
Average number per sample	9·6	22·0	68·8	100·4

How do your results compare with those in the table? Were you more successful in scooping out nearly 100 beans each time than the authors?

It may have surprised you a little to find how close the average sample sizes are to the expected sample sizes, even though we did not know exactly how large the whole population was, and even though the samples were not all the same size.

It is because the technique of sampling can overcome such difficulties that it is so powerful. There are, however, precautions that we can take to make the whole method even more reliable, and we shall now examine two of the most important.

Experiment 2. The size of the sample

Method. Carry out the previous experiment, but use a sample of 10 beans each time. These can be selected either by using a small scoop such as a bottle-top, or by using your fingers, but it is important in such cases to select each sample with shut eyes, as it is very easy to cheat unintentionally.

Record your results in a table as before, first working out the totals and averages after twenty-five samples have been taken, and then going on to take fifty samples in all. Work out the results again for this number.

Discuss your conclusions with the rest of the class; why were twenty-five samples suggested this time, and not just ten as in the previous experiment?

It should become clear that it is quicker to take a small sample, but the smaller each sample, the more samples we need to take before we can be certain of our results. Also, the smaller the sample, the more important it is to make sure that each sample is the same size—two or three samples of perhaps 15 beans when all the others are exactly 10 beans may have a great effect on the final results.

The second precaution that we must take in sampling is to ensure that we always have what is called a RANDOM sample. This is why we have been careful always to stir the contents of the basin before sampling. We shall now discover what can happen if we do not ensure randomness in our samples.

Experiment 3. A sample must be random
Sort all the beans used in the previous experiment into three heaps, red, blue and white. Place the red beans into the basin, pour the blue beans carefully on top of the red, and then cover the blue beans with the white ones. DO NOT MIX the whole population at all.
Method. Again fill your matchbox scoop, but take care to scoop your sample from the top without disturbing the rest of the beans.

Sort each sample you take by colours and record in a table as before. Return to the basin and repeat the process, but without deliberately mixing the whole population at any time.

A sample must be random

How do these results compare with either the expected results or with the results of the first experiment?

It is obvious what is wrong: the sample we have taken each time is in no way typical of the contents of the whole basin, but only the top layer. By not stirring and by taking only a shallow scoop we have not obtained a random sample. It is most important that a sample should be random, or typical, if we are to place any reliance in the results.

Exercise A

Do you think the following sampling methods would be random? Give reasons for your answers, and where not random suggest better methods:

1. Determining the most popular brand of cigarette by questioning passengers in a non-smoking compartment of a train.

2. Determining the number of TV sets in your town by asking 100 people outside the local cinema whether they have a TV set or not.

3. Determining the most popular make of car by counting 500 cars at a roundabout.

4. Selecting a form representative by picking a name out of a hat.

5. Deciding whether potatoes are cooked by prodding one potato with a fork.

Randomness

We have seen that it is most important to ensure that any sample we take is in fact random. It is not always easy to achieve this as the following example will show.

Class experiment

Every member of the class should write down, without consultation, a number from 1 to 9 on a piece of paper. Record the results on the blackboard or on the class bar chart.

There seems to be every reason to expect the final distribution for the whole class to be random, that is, each number should be selected as often as any other number. Another way of looking at this is to say that such a distribution ought to be rectangular.

Are the one-digit samples of each pupil from the total population of nine digits truly random, or has some bias affected the results?

The outcome of this experiment cannot really be explained—it is just one of those things—but it is just such things which can affect the whole technique of sampling. Such considerations become very important in Gallup Poll surveys: personal bias is very hard to eradicate because it is so hard to predict or avoid. We must nevertheless recognise that it does exist, and try to devise sampling methods which will eliminate personal feelings.

In order to achieve this impersonal approach we often use a method known as SAMPLING BY RANDOM NUMBERS.

A set of 20-faced dice produced by the Japanese Standards Association to generate random numbers.

Suppose we wish to select five members of your class to form a typical sample in some larger survey within the whole school. If we merely left the selection to a vote, personal preferences would creep in, for example the most popular girl might well be selected although she might be very untypical of the whole class in many ways. If indeed there happened to be only one girl in a class of boys, that girl would very likely be voted in. This would certainly create a wrong impression as far as the proportion of boys to girls was concerned, as it might be deduced that the class had one girl to every four boys, which would not be accurate.

She might be very untypical of the whole class

Sampling by random numbers consists in assigning a number to every member of the class, and then consulting what are called TABLES OF RANDOM NUMBERS in order to select five numbers. These numbers determine the actual five members of the class chosen to form the sample.

Exercise B

1. Throw a die 50 times, recording the results in order.

Do you think this is a table of random numbers?

Can you predict what the 51st number would be?

Draw a bar chart of the whole distribution: does this confirm your previous conclusion?

2. Will the loaded die you made in Chapter 4, p. 43, produce a list of random numbers? Try it and see if you can recognise a pattern. Draw a bar chart of 50 results.

3. Do you think the numbers 1, 3, 5, 6, 4, 2, 1, 3, 5, 6, 4, 2, ... are a list of random numbers, or can you see a pattern which will enable you to write down more numbers of the same sequence? Draw a bar chart of the frequencies of the first 30 numbers of this list; is it true that a rectangular distribution must be a random distribution?

4. Do you think the following methods will generate lists of random numbers? Try them, and see whether all possible numbers appear not only equally frequently, but in a completely unpredictable order:

a. Take a book of about 300 pages and open it at random 25 times. Record each time:

   i. The righthand page number.

   ii. The digit-sum of the lefthand page number.

   iii. The final digit of the righthand page number.

   iv. The sum of the two page numbers.

Explain your conclusions in each case.

b. Cut a pack of cards, and record the face value of the card revealed, writing 0 whenever a 10, J, Q, or K appears. Repeat the process but writing 0 whenever a 10 appears and ignoring any appearance of J, Q or K.

Explain your conclusions in each case.

Use the card method to generate a list of 50 random single-digit numbers.

Write these down in the order in which they were generated, and pair them together in the order in which they have been written down, and so generate a list of 25 random two-digit numbers.

5. Suppose there are 34 pupils in your class.

Number the members 01, 02, 03, . . ., 33, 34, in alphabetical order.

How could you use the list of 25 two-digit random numbers which you have produced to select a random sample of five members of your class?

It is probably simplest to strike out all two-digit numbers in your list which are greater than 34. If every member of the class selects his own class sample in this way, each member of the class should, on the whole, appear in an equal number of samples.

Does this in fact happen?

6. Invent methods by which car registration numbers, or telephone numbers, could be used to generate lists of random numbers.

Put your method into practice, and test the final list for randomness.

7. Here is the value of π to 206 figures; is the sequence of numbers random?

$$\pi = 3 \cdot 14159265358979323846264338327950288419716939937510582097$$
$$49445923078164062862089986280348253421170679821480865132$$
$$82306647093844609550582231725359408128481117450284102701$$
$$9385211055596446229489549303819644 2881 \ldots$$

Class projects

Find out how information is obtained in the following real-life surveys.

Obtain and analyse some actual figures in each case, using the techniques you have learnt. How is randomness achieved in the different examples?

What conclusions can you draw from these investigations?
Display your results for others to see.

a. Television audience measurement.

b. Top-of-the-pops ranking.

c. Design of bingo cards.

d. Gallup poll election forecasting.

e. Selection of premium bond winners.

Apparatus

As you have seen, a great deal of useful and interesting work can be done with the simplest of apparatus such as dice, cards and beans. However, some pupils may have the ability and facilities to make more advanced apparatus and details are given here.

Sampling

If a large number of ball-bearings can be obtained, sampling as in the bean experiment can be improved.

Ball-bearings can easily be coloured in two ways:

a. Dip them in copper sulphate solution and they become copper coloured.

b. Heat them up to red heat and plunge them into oil, when they will become blue coloured.

These, together with the untreated balls take the place of the red, blue and white beans with the following advantages:

a. They mix much better.

b. Special ladles can be made so that the composition of the sample can be seen immediately. These are simply rectangular pieces of wood, metal or perspex with handles, and holes drilled part of the way through. Push this into your box of bearings and shake it until all spare ones fall off leaving one in each of the holes.

This is a very quick way of getting lots of samples.

Dimensions will depend upon the size of ball-bearing you have been able to obtain. A set of different ladles can be made for different size of sample.

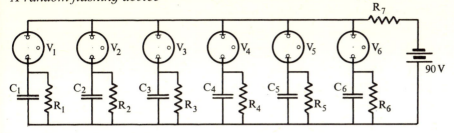

V 1 - 6 Wire-ended neons.   R 1 - 6   10 MΩ $^1/_2$ W Resistors

R 7   2MΩ $^1/_2$ W Resistor

C 1 - 6   0·47 μF 250 V Wkg. paper capacitors.

90 V radio battery (Ever Ready B 126)

This device has six neons which flash at varying intervals and in a seemingly random fashion. This can be quickly and easily made at a cost of about £1.

After plugging in, leave the circuit for a few minutes to allow it to settle down. The randomness depends on the tolerances of the components and consequently so does the rate of flashing. If it is difficult to observe the actual order of flashes because they occur too frequently, try changing the value of R7.

Although there is apparent randomness, it is never strictly random in the mathematical sense. See if you can carry out an experiment to show that this is so.

An electric die

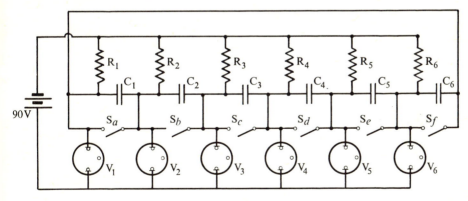

V 1 - 6   Wire-ended neons.   R 1 - 6   2MΩ $^1/_2$ W Resistors

C 1 - 6 0·1 μF 250 V Wkg. paper capacitors.

S a - f Six-pole two-way wafer switch, or keyswitch

90 V radio battery (Ever Ready B 126)

Many of the components from the random flasher can be reused to make the electrical die; this is basically the same circuit with the addition of a six-pole two-way switch, and different capacitors, which will cost about £0·50 extra.

The neons flash—in this case much faster than before—and when the switch is closed the one which was flashing at that instant is held alight.

Test this device for randomness. Is there any obvious difference between the results of the random flasher and the electrical die?

8 Spread

DULL WITH SHOWERY PERIODS

MIN: 15°C
MAX: 21°C

You have probably noticed that television weather forecasts do not often give the average temperature for the next day, but usually give the maximum and minimum temperatures expected. This is done partly because a truly average or typical temperature would be hard to estimate, but chiefly because the two predictions, an upper and a lower, are the measures of temperature which are of most value to the person listening to the forecast.

People are not concerned with the actual temperature at noon, or the temperature which will occur most frequently during the day, but rather with how hot and how cold it will be. This is the information we require when we decide whether or not we need a coat, whether we ought to adjust the thermostat on the central heating, or whether we can plant out our seedlings without them being either frozen or scorched.

Such a measure of a set of figures is called the RANGE or SPREAD; it is frequently used to supplement the information given by the average value of a set of numbers.

Class projects

1. The following table gives the daily range of temperatures for a week in April in London.

	Sun.	Mon.	Tues.	Wed.	Thur.	Fri.	Sat.
Maximum °C	11	8	15	14	7	$9\frac{1}{2}$	9
Minimum °C	2	7	9	9	5	5	$2\frac{1}{2}$

Plot these results on graph paper, putting a red cross for each maximum entry and a blue cross for each minimum entry.

2. Set up a maximum and minimum thermometer, and from it obtain the daily range of temperature at your school from Monday to Friday in any one week. The class will have to decide the most suitable location for the thermometer: if it is inside, it may be affected by central heating—if it is outside, it may need protection from the weather or from other people.

Record your results on graph paper and compare the figures with those published in a newspaper for the same days.

3. Obtain similar figures from a different part of the country, and compare your results with these. (Your public library might have copies of newspapers published in other large cities, or someone in the class may have a friend living in another part of the country who might be able to send the figures.)

It would be interesting to compare your figures with those from an inland town if your school is near the coast, or with a place in the north of Scotland if your are living in the south of England. Do inland places on the whole have a greater range of temperature than places near the sea?

The spread of a set of numbers is a measure not so much of the actual size of some quantity, but of how variable that quantity may be.

Consider the following example. Every week a class of pupils is given a mathematics test. The results over a ten-week period for two pupils are:

Test number	1	2	3	4	5	6	7	8	9	10
Maximum possible	20	20	20	20	20	20	20	20	20	20
Anne Wakefield	19	15	9	2	11	9	10	12	4	18
Jane Nixon	12	11	10	12	12	10	11	11	10	11

What comments can we make on these marks?

If we add up the marks awarded to each girl, we find that the two totals are the same:

	Total	Average/test
Anne	110	11
Jane	110	11

The average marks, by themselves, lead us to the conclusion that both girls are of the same standard in mathematics. However, a closer look at the actual scores suggests that the performances of these two are somewhat different.

Anne is obviously much more variable than Jane. This can be seen very clearly if the two sets of marks are plotted on graph paper, but it would be very useful if we could invent some measure of the variability.

Jane's lowest mark is 10 and her highest is 12; she has a range of only 2 marks. Anne's lowest mark is 2 and her highest is 19; her range is 17. The range, or spread, of the two sets of marks gives us an easy and quick numerical way of assessing the amount of variability.

Variability is a very important feature in any assessment of numerical information. A pupil with a small range of marks might not be very good, but will at least be reliable and will produce consistent results. A pupil with a wide range of marks is not very reliable and will have good days and bad days. Such variability is often an indication of a careless approach to work.

The same considerations are very important in industry. A machine, or a typist, producing very variable work is more of a liability in many ways than one with a reliable output.

The range is a measure, not of actual output, but of reliability. It has the great advantage that it is very easy to work out, being the difference between the largest and smallest values, but it must be realised that this is only a very rough guide of reliability.

The following table shows the numbers of rejects per day from three machines producing electrical components in a factory:

	Mon.	Tues.	Wed.	Thur.	Fri.	Total	Average	Range
Machine A	35	26	28	35	26	150	30	9
Machine B	27	33	32	27	31	150	30	6
Machine C	28	28	28	28	38	150	30	10

The firm's efficiency expert naturally wants to know which machine has the most reliable output.

In this particular case, each machine has the same weekly output of rejects. Judged only by its range, machine C is the least reliable, since it has a spread of 10, but look at its reject numbers day by day. Until Friday, it was working most consistently—perhaps there is some special reason for its failure on Friday, like an inexperienced operator? Here we have an example of a machine capable of producing the most consistent results, and yet the range alone would condemn it as being the most unreliable.

The calculation of the spread alone is insufficient to establish such details, and in practice a works-study engineer can use more advanced ways of measuring variability: one measure is called the MEAN DEVIATION and is calculated like this:

Machine A	Number of rejects	Deviation from mean of 30
Monday	35	5
Tuesday	26	4
Wednesday	28	2
Thursday	35	5
Friday	26	4
Totals	150	20
Mean	$\frac{150}{5} = 30$	Mean deviation $\frac{20}{5} = 4$

The separate steps in the calculation are:

a. Find the mean number rejected each day (by any of the methods given in Chapter 2, pages 20, 23–24).

b. Find by how much the total for each day differs from this mean—this is called the DEVIATION.

c. Find the mean of all the deviations.

The mean deviation is the average of all the differences between the actual mean value of 30 and the individual figures for each day. Notice that no account has been taken of plus and minus signs in the calculation; a value of 33 which is 3 above the average contributes a deviation of 3 to the mean deviation. A value of 27 is 3 below the average but also contributes a deviation of 3 to the mean deviation.

Can you think why we must ignore the plus and minus signs in this calculation? What happens if we do include the signs and take the contribution of 35 to be $+5$, the contribution of 26 to be -4, and so on?

Here is the same calculation for machine C:

Make sure you understand all the details of the calculation.

Notice the use of the symbol $\sim$ which is sometimes read as 'twiddle'. $x \sim m$ is the difference between x and m irrespective of which is the larger.

Calculate the mean deviation for machine B and now see if you can decide which of the three machines is the most, and which is the least, variable using mean deviation as a measure.

	Number of rejects	Deviation from mean
	x	$x \sim m$
Monday	28	2
Tuesday	28	2
Wednesday	28	2
Thursday	28	2
Friday	38	8
Totals	150	16
Mean (m)	30	3·2

1. Here are the marks for a group of pupils in 10 weekly French tests. Give the range of marks for each pupil:

Week number	1	2	3	4	5	6	7	8	9	10
Joan	9	10	10	6	8	8	10	9	5	8
Margaret	8	8	7	8	8	7	8	9	8	8
Ann	4	3	4	4	5	4	5	4	3	3
Helen	6	6	8	6	6	6	8	7	7	7

Calculate the mean mark and the mean deviation from the mean for each pupil.

Who is the best at French? Who is the most variable?

2. The following are the numbers of red beans obtained in ten samples by three different pupils when performing the bean experiment of Chapter 7, page 72.

Pupil A	8	12	13	6	9	10	11	14	9	9
Pupil B	14	8	12	9	7	3	10	12	11	12
Pupil C	15	7	6	7	12	11	9	9	13	11

What will be the estimates of the mean number of red beans in each case?

What is the mean deviation in each case?

Which is the most reliable set of results?

3. Calculate the mean, the range, and the mean deviation for the following set of numbers:

a. 4, 6, 8, 10, 12, 14, 16, 18, 20, 22. c. 4, 4, 4, 4, 4, 22, 22, 22, 22, 22.
b. 4, 4, 10, 10, 10, 16, 16, 16, 22, 22. d. 4, 12, 12, 12, 12, 14, 14, 14, 14, 22.
e. 4, 6, 6, 6, 12, 14, 20, 20, 20, 22.

4. Express the marks you have scored in your last five tests in any one subject as marks out of 100.

Calculate your average percentage mark for this subject over five tests, and estimate your mean deviation for these tests. Draw up a list which ranks the members of your class in order of: (a) average marks and (b) mean deviation.

Is the pupil with the highest average mark necessarily the most reliable pupil?

9 Continuity

There is no doubt that civilisation has made progress in the past century. You can probably think of many domestic improvements but there is one important difference in the illustration which has great mathematical significance. How are the two baths filled?

There can be no doubt that it is much easier to turn a tap than it is to carry bucket after bucket from the kitchen stove; the difference in the two methods is even more marked if the quantity of water in each bath is represented graphically. Suppose each bath holds 100 litres and each is filled in 5 minutes, the tin bath with a 10 litre bucket, the modern bath with a tap.

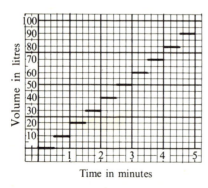

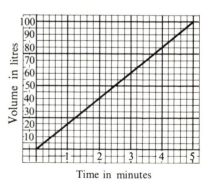

Mathematically we say that the tin bath is filled in DISCRETE steps, the modern bath by a CONTINUOUS process.

Consider the following situations in which two similar processes are described. Which are the discrete, and which are the continuous? In some of the cases, more discussion may be necessary before a clear decision can be made.

a. Deepening a harbour entrance by using a bucket dredge or a suction dredge.

b. Inflating a tyre by using an air line or a hand pump.

c. Growing a tree from a seedling or building a house brick by brick.

d. Going up an escalator or walking up stairs.

A formal definition of the two words 'discrete' and 'continuous' is not necessary at this stage, but it is most important that you are clear in your own mind about what is involved, as later mathematical methods sometimes depend upon recognising whether a particular process is discrete or not.

It is sufficient for the moment to say that the graph of a continuous process has no *jumps*, while the graph of a discontinuous or discrete process may even consist of a set of isolated points.

It is interesting to realise that what may appear to be continuous in one comparison may be distinctly jerky in another context. Consider the cinéfilm coverage of a certain event, say the Cup Final, and a succession of stills of the same match. At first sight the cinéfilm would be called continuous, but in fact it is nothing more than a succession of discrete stills very closely spaced.

A CONTINUOUS VARIABLE is one which is capable of taking any possible value from a certain set, while a DISCRETE VARIABLE is one in which the possible values are restricted.

For example, the quantity of water in the luxury bath can, theoretically, be any value between 0 litres and 100 litres during the 5 minutes of the filling. This is a continuous variable. On the other hand, the quantity of water in the tin bath can only take certain limited values between 0 and 100 litres during the same time—we cannot really estimate the time at which the tin bath holds 55 litres.

This idea of continuity helps when we are drawing graphs. If a variable is known to be continuous, then we can join up the plotted points of a relationship; if the variable is discrete, however, the separate points must be left separate. Joining up the points of a discrete graph may give quite the wrong impression.

Example

In the Smith family, two important happenings take place at the end of each half-year. The height of Jane Smith, the daughter of the household, is recorded against the bathroom door, and Mr Smith receives in cash a small rebate from the local council because he always settles his rates promptly. Over a four-year period, the results of the two happenings are:

Date	June 66	Dec. 66	June 67	Dec. 67	June 68	Dec. 68	June 69	Dec. 69
Height of Jane in cm (x)	102	105	108	110	112	113	114	115
Cash rebate for Mr Smith in pence (r)	102	105	108	110	112	113	114	115

These two tables can be plotted as shown:

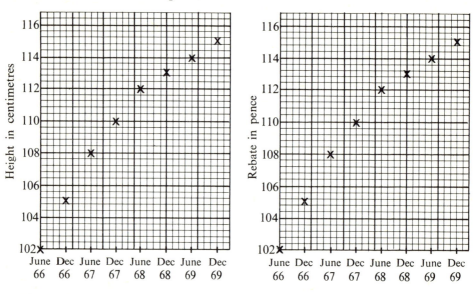

The question which then follows is whether or not the points can be connected. Undoubtedly a TREND LINE is clear in both graphs—both x and r show a tendency to increase each half-year—but only the points on the x graph can be connected by a smooth line. It is only the height of Jane which has any meaning between the actual values already plotted.

It is reasonable to deduce that Jane's height in September 1967 was about 109 cm; it is nonsense to claim that Mr Smith's rebate during the same month ought to be 109p, as he cannot receive *any* rebate except at the end of each six-monthly period.

The marks we make on our graphs must always correspond to events which can actually happen, and a non-continuous variable must not be made to look continuous by joining up points which are discrete.

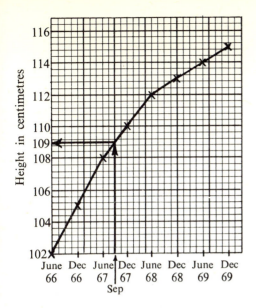

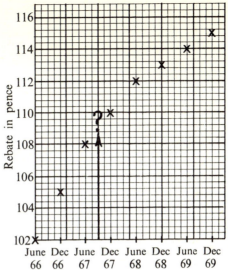

Exercises

Draw graphs to illustrate the following tables. In each case decide whether the variables are continuous or discrete, and whether or not further readings could be obtained from your graph.

1. Reading on car speedometer in km	13,756	13,856	13,926	13,986	14,036
Reading of petrol gauge	full	$\frac{3}{4}$	$\frac{1}{2}$	$\frac{1}{4}$	empty

What factors do you think might account for this graph not being a straight line?

2. Number of passengers in a bus	15	12	7	8	10	12
Fare stage number	0	1	2	3	4	5

Can you say at what part of the journey the bus had 11 passengers on board?

If each fare stage is 1 mile in length, can you say how many passenger-miles have been completed in this journey?

If the fare is 3p per fare stage, what is the total fare taken by the conductor during the journey?

89

3. The depth of water at a harbour entrance during a certain day:

Time of day (T)	6 a.m.	7 a.m.	8 a.m.	9 a.m.	10 a.m.	11 a.m.
Depth of water in metres (d)	5·6	7·2	8·8	10·1	11·0	11·6

Time of day (T)	12 noon	1 p.m.	2 p.m.	3 p.m.	4 p.m.	5 p.m.	6 p.m.
Depth of water in metres (d)	11·7	11·2	10·5	9·2	7·8	6·1	4·4

Estimate the time of highwater on this particular day.

Between what times is it safe for a ship drawing 9 m to enter or leave the harbour on this day?

4. The height of a plant in inches:

Number of weeks since planting (n)	1	2	3	4	5	6	7
Height in cm (x)	0·5	12·4	22·1	26·2	27·9	29·4	30·2

These plants should be transplanted when they are 15 cm high. When should this happen?

5. The amount of pocket money John Brown has left at bedtime each night:

Day	Fri.	Sat.	Sun.	Mon.	Tues.	Wed.	Thurs.
Pocket money left	25p	17p	7p	$5\frac{1}{2}$p	$3\frac{1}{2}$p	2p	$\frac{1}{2}$p

On what day does he spend his tenth penny?

If he goes to bed regularly at 10 p.m. each evening, is it possible to estimate accurately at what time on Wednesday he has only $2\frac{1}{2}$p left?

10 Further averages

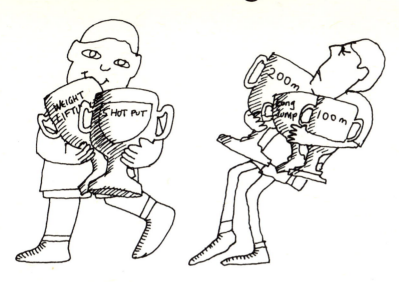

Weighted average

Two pupils in the fourth form found that they had the following marks in the end of term examinations:

	Pupil A %	Pupil B %
English	62	53
Mathematics	28	69
Science	30	74
History	70	37
Geography	66	36
Art	80	31

Pupil A has a total of 336 marks and pupil B has a total of 300. Both took six subjects, so their averages are:

A, 56 marks;   B, 50 marks.

From this point of view it appears that A is doing better than B, and in the general, overall sense, this is true.

However, what if both pupils were hoping to make their careers in science or technology? B got high marks in Mathematics and Science whereas A got low marks in these subjects. From this standpoint it could be argued that B's performance was better than A's. The subjects B did poorly in were not very important to his career anyway.

There is an easy way of making an allowance for this situation. First of all decide on the relative importance of each subject. Let us suppose that the following order is decided upon:

Mathematics⎫
Science   ⎭ Equal first in importance.

English   Next in importance.

History ⎫
Geography⎭ Equal next in importance.

Art   Least important.

Start at the bottom with art, and give this a 'weighting' of 1. Give history and geography each a weighting of 2, English a weighting of 3 and mathematics and science top weighting of 4.

Set out the following tables:

Pupil A	Marks (%) x	Weight w	$x \times w$
English	62	3	186
Mathematics	28	4	112
Science	30	4	120
History	70	2	140
Geography	66	2	132
Art	80	1	80
Totals		16	770
Weighted average		$\frac{770}{16} = 48\frac{1}{8}$	

Pupil B	Marks (%) x	Weight w	$x \times w$
English	53	3	159
Mathematics	69	4	276
Science	74	4	296
History	37	2	74
Geography	36	2	72
Art	31	1	31
Totals		16	908
Weighted average		$\frac{908}{16} = 56\frac{3}{4}$	

Notice how the calculation was carried out; a column for marks multiplied by weight was included, and the total of this column was divided by the total number of weights.

Summarising all the results we have:

	Pupil A	Pupil B
Simple average	56	50
Weighted average	$48\frac{1}{8}$	$56\frac{3}{4}$

This now gives **B** a better average mark than **A** and reflects the fact that he did well in his important subjects.

The result depends upon the weighting which is given to the subjects. In different circumstances it might be decided to weight these six subjects quite differently.

Moving average

A car dealer records the total number of cars sold every quarter over a period of three years, and obtains the following figures:

	1967	1968	1969
1st quarter	6	10	9
2nd quarter	16	16	18
3rd quarter	25	30	32
4th quarter	11	7	10

Plotted on a graph, this will give:

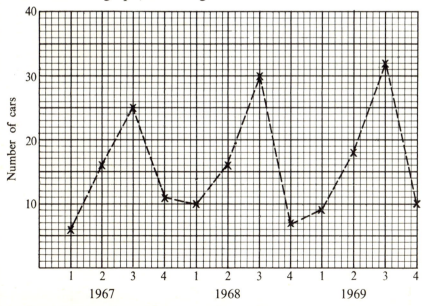

Sales have soared up and down, but this is quite normal. People tend to buy new cars at certain times of the year, and the graph shows a SEASONAL

FLUCTUATION. A series of values over a period of time is called a TIME SERIES, and this is a time-series graph.

The car dealer is not so interested in the seasonal fluctuation as in the *general trend* of his sales. Is business improving? It seems as though it is a little, and we shall now devise a way to show this graphically.

Find the average for 1967: $\quad \dfrac{6+16+25+11}{4} = 14\tfrac{1}{2}$

Next the average for the 2nd, 3rd and 4th quarters of 1967 and the 1st quarter of 1968: $\quad \dfrac{16+25+11+10}{4} = 15\tfrac{1}{2}$

Next the average for the 3rd and 4th quarters of 1967 and the 1st and 2nd quarters of 1968: $\quad \dfrac{25+11+10+16}{4} = 15\tfrac{1}{2}$

Continue, moving the average along each time by dropping off one quarter and taking on the next until you reach the average for 1969:

$$\dfrac{9+18+32+10}{4} = 17\tfrac{1}{4}$$

These results are best set out in a table:

	1967				1968				1969			
Quarterly sales 6	16	25	11	10	16	30	7	9	18	32	10	
Four-quarterly moving average		$14\tfrac{1}{2}$	$15\tfrac{1}{2}$	$15\tfrac{1}{2}$	$16\tfrac{3}{4}$	$15\tfrac{3}{4}$	$15\tfrac{1}{2}$	16	$16\tfrac{1}{2}$	$17\tfrac{1}{4}$		

If these values are now plotted on the previous graph and joined up, we get:

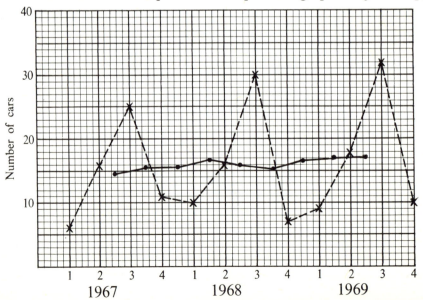

Notice how much smoother the new graph is, and how it shows a slight upward trend. Notice also that the points have been plotted in the middle of the interval for which they are the average.

It is easy to take a short cut when calculating moving averages. Consider the car sales figures:

The first average was: $\frac{1}{4}(6+16+25+11) = 14\frac{1}{2}$

The next one was: $\frac{1}{4}(16+25+11+10)$.

Notice how this differs from the first one: 6 has been dropped off one end and 10 taken on at the other end. This really amounts to $10-6$, that is, 4 more cars and when divided by 4 gives one extra to the average which then becomes $15\frac{1}{2}$.

The next one is: $\frac{1}{4}(25+11+10+16)$,

16 has been dropped from the beginning, and 16 added on. The average remains the same as before, namely, $15\frac{1}{2}$.

The next time we drop 25 and take on 30, so 5 more cars are added. This will increase the average by $1\frac{1}{4}$ and gives $16\frac{3}{4}$ and so on.

Exercises

1. This table gives the marks awarded to four candidates in an examination.

		Candidate			
Subject	*Weight*	*A*	*B*	*C*	*D*
English	3	52	71	36	47
French	1	61	65	31	52
Mathematics	2	47	21	74	60
Geography	1	65	47	57	52
History	1	39	38	60	63
Science	2	23	16	70	55

Calculate:

   a. The simple average for each candidate.

   b. The weighted average for each candidate.

   2. In a first-year school test Alison scored the following percentage marks: mathematics 80, science 93, English 73, geography 24, history 35, handwork 27, music 38, and religious instruction 10.

Her friend Ben obtained the following marks (given in the same order as Alison's): 39, 64, 27, 42, 81, 74, 33 and 35.

The Headmaster always gives treble weight to mathematics and English and double weight to science. Find the children's weighted mean percentage and thus work out whether or not Alison did better than Ben.　(EA)

3. The following percentage marks were obtained by a pupil in a certain examination. Mathematics 75, English 73, French 52, physics 82, geography 64, history 68.

a. Find the arithmetic mean of these marks.

b. If mathematics and English are given treble weight and French and physics double weight, find the weighted mean of the marks.

c. Why is the weighted mean used rather than the arithmetic mean? (SE)

4. Record the number of absentees in your school every day over a period of four consecutive weeks. Make a time-series graph of this data and comment upon any fluctuation.

5. The figures below show the amounts (in millions of pounds) spent on food during twelve consecutive quarters:

	First year	*Second year*	*Third year*
First quarter	714	790	876
Second quarter	768	841	936
Third quarter	814	876	953
Fourth quarter	833	909	966

a. Draw a graph to illustrate the data given in the table above.

b. Calculate and tabulate the four-quarterly moving averages.

c. Draw a graph of the four-quarterly moving averages.

d. Comment critically on these graphs. (MX)

6. The following are the gross sales figures of a certain sales team:

Month	£	*3-monthly moving totals*	*3-monthly moving average*
January	150		
February	180		
March	200		
April	190		
May	305		
June	320		
July	240		
August	265		
September	280		
October	315		
November	290		
December	265		

a. Complete the last two columns of the table.

b. Plot the original sales figures as a line graph.

c. Plot the 3-monthly moving averages on the same axes.

d. Give the advantages of plotting the moving averages. (SE)

11 Diagrams

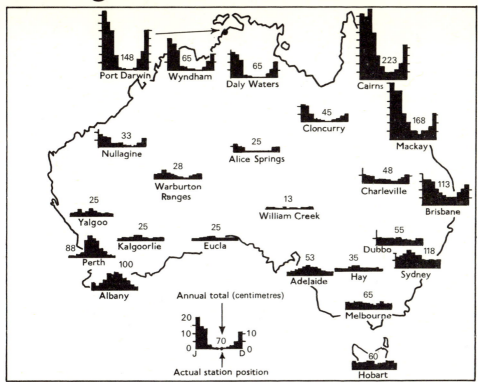

As we saw in chapter 1, the display of statistical information is very important; in fact whole books exist on this topic alone. In this chapter we shall have a further look at the presentation of data by graphical methods.

Often there is no absolute right or wrong, some methods are undoubtedly better than others, and the diagram depends upon the ingenuity of the person producing it. However, there are some rules to be followed and the examples chosen will help to bring these out. With all diagrams we must be especially careful, for it is easily possible to deceive people by distorting a graph, or by omitting some essential piece of information.

The histogram

Consider the following problem: these figures give the number of deaths of children under five years old in a certain town during one year. Represent this data graphically.

Ages in years	0–1	1–2	2–3	3–5
Number of deaths	5	4	4	6

We *might* draw this diagram:

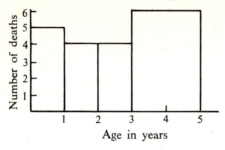

Age in years

but it would be very misleading as it appears that the worst group is the last one. In fact the six recorded deaths were over *two years*, all the other entries were for *one* year only.

It would be better if we assumed that the last group had three deaths in each year, giving a total of six for the two years. Our diagram would now be:

This is still not entirely accurate. (Why?) But it gives a much better representation of the facts. Such a diagram is called a HISTOGRAM.

When drawing a histogram we must be careful to inspect the class intervals and see if they are all the same; if not, adjustments must be made.

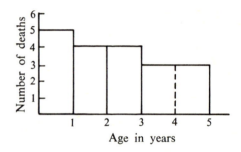

Age in years

Example

Represent this population distribution as a histogram:

Age group	0–4	5–9	10–14	15–24	25–34	35–54	55–64	65–74
Popn '000s	4·0	3·5	3·2	6·0	5·8	10·4	3·6	1·4

The first thing to notice is that 0–4 means from day of birth until the day before the fifth birthday; that is, a *five* year span.

The second thing to notice is that the class

Age group	Population '000s	Interval (years)	Height of column (units)
0–4	4·0	5	4·0
5–9	3·5	5	3·5
10–14	3·2	5	3·2
15–24	6·0	10	3·0
25–34	5·8	10	2·9
35–54	10·4	20	2·6
55–64	3·6	10	1·8
65–74	1·4	10	0·7

When the third column, 'Interval', has been entered, it will be seen that 5 years can be taken as the unit interval. The age group 15–34 spans two units of 5 and the population of 6·0 thousands is split up into two parts, each of 3·0 thousands.

The histogram will appear as:

Notice that with many bar charts, the bars can be moved together so that they are touching, and they can then be called histograms.

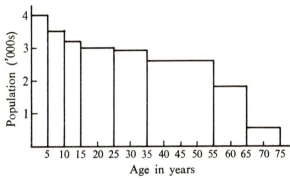

Frequency polygon

If a distribution is first represented as a histogram and then the middle points of the tops of each column joined, the resulting figure is called a FREQUENCY POLYGON.

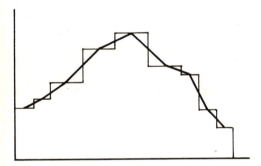

Other diagrams

1. The diagram at the beginning of the chapter shows the distribution of rainfall in Australia by putting little bar charts showing the annual distribution month by month at many places all over the country. This gives an enormous amount of general information at a glance; the details of each can be found fairly accurately by reference to the key at the bottom of the diagram.

2. A common way of comparing data is to put two bar charts next to one another, like the one on the next page. Notice also the symbols at the bottom of each bar which make the diagram more eye-catching.

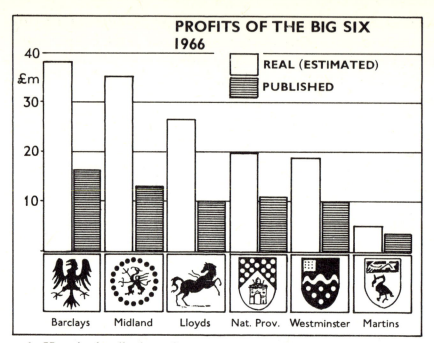

PROFITS OF THE BIG SIX
1966

REAL (ESTIMATED)
PUBLISHED

Barclays   Midland   Lloyds   Nat. Prov.   Westminster   Martins

3. Here is the display of a survey concerning the arrival and departure of Southern Region trains in London. The two charts are placed side by side so that it is easy to notice the difference in pattern. It would be wrong to superimpose one on the other since different times of day are involved.

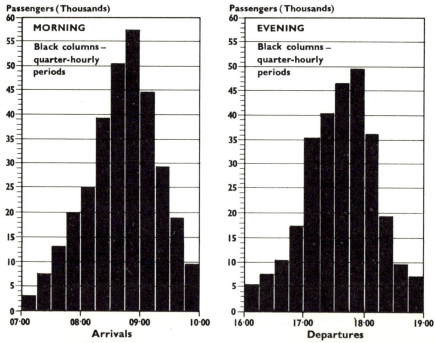

Passengers (Thousands)

MORNING

Black columns — quarter-hourly periods

Arrivals

Passengers (Thousands)

EVENING

Black columns — quarter-hourly periods

Departures

4. This is a most unusual and interesting form of bar chart where the bars are really distances along a road. It is not very plain to see at a glance, but very meaningful to a motorist who knows the area.

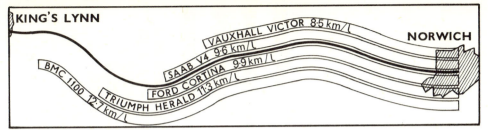

5. Another form of comparison is given by this pair (*a*) of sideways bar charts. Upright bar charts can be used for this sort of comparison by having one above, and one below a line (*b*).

(*a*)

(*b*)

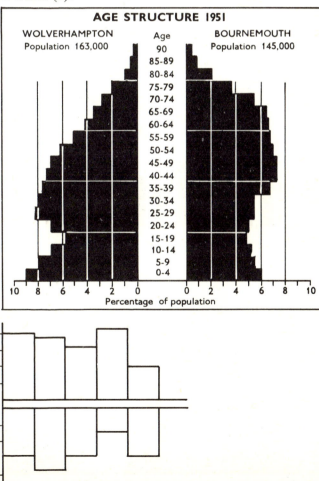

6. Why do the spaces differ in between the bars?

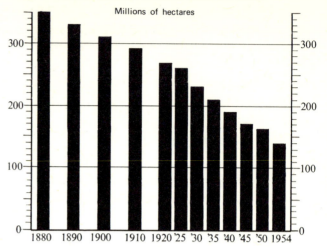

Millions of hectares

7. There is exactly the same data in both charts. Do you prefer one to the other?

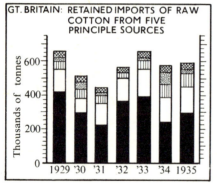

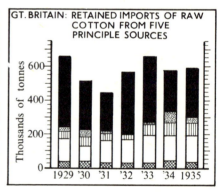

GT. BRITAIN: RETAINED IMPORTS OF RAW COTTON FROM FIVE PRINCIPLE SOURCES

KEY

U.S.A.

Brazil

India

Egypt

Peru

8. An attempt to represent data by comparing the *areas* of circles. Is it obvious that the number of requests for home helps is about twice the number of requests for health visitors?

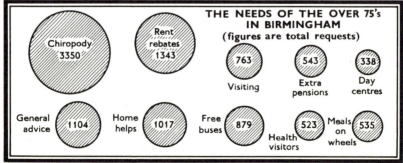

THE NEEDS OF THE OVER 75's IN BIRMINGHAM
(figures are total requests)

The surprising results of Birmingham's Welfare survey

9. This circular bar chart represents the mean monthly rainfall in Athens over a period of years. It has been bent into a circle because this information does not have a beginning or an end; we want January's figure to follow immediately after December's. This is not the picture for one special twelve months, but the typical picture for any year.

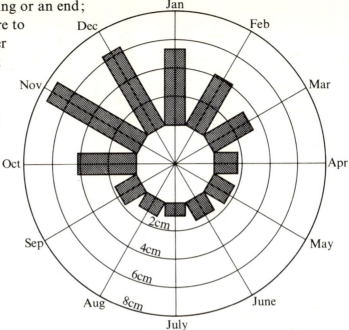

10. The line graphs showing the growth rates of the Rio Tinto Zinc concerns; the inclusion of the map gives the geographical location.

Here are the figures for Australia; portray them similarly:

	1963	1964	1965	1966
Millions £	41·9	47·6	57·4	89·4

Notice that there are four values plotted for each line in Africa. What important information is missing from the African graph, but can be included in the Australian graph?

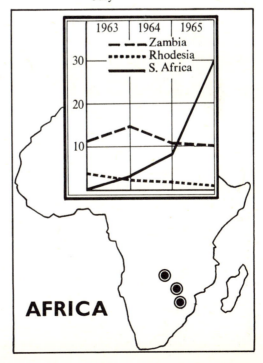

11. Occasionally a line graph can be improved by shading, but be careful with line graphs; remember what you learnt about continuity in Chapter 9.

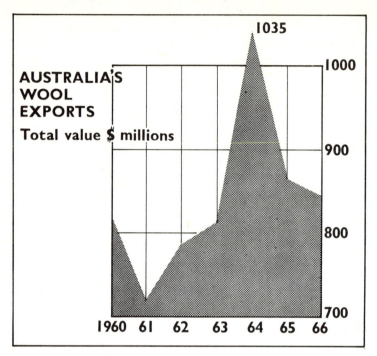

12. Attempts to put many variables on one graph can soon make it too complicated, and the advantages of graphical representation are reduced.

1. This table gives the mean monthly rainfall for Perth in Australia.

Month	Jan.	Feb.	Mar.	Apr.	May	June	July	Aug.	Sept.	Oct.	Nov.	Dec.
Rainfall cm	1·0	1·3	2·0	4·3	13·0	17·8	17·3	14·5	8·9	5·6	2·0	1·5

Represent this on a circular bar chart.

2. Here are three postage stamps which portray graphs:

Population census.

Population census.

Eradication of malaria.

Say what you think is wrong with each one.

3. The number of goals scored by each of the twenty-two teams in the First Division of the Football League in the season 1965/66 is given below:

79, 72, 79, 84, 65, 91, 80, 75, 56, 65, 56, 70, 55, 62, 50, 69, 56, 56, 51, 67, 55, 57

Tabulate these figures in a frequency distribution, starting at 50 and taking class intervals of 5.

Use this distribution to produce a histogram showing the information. (AL)

4. The following marks were obtained by a group of 40 pupils in an examination:

```
55   15   40   35   60   40   20   35
30   35   40   35   50   25   40   30
55   40   55   35   65   60   35   40
30   35   25   35   55   30   50   25
20   30   55   30   35   60   50   55
```

Tally these marks in a frequency distribution, and draw a frequency polygon.

The results of the examination were given in the following categories:

0–35 fail; 36–55 pass; Over 55 pass with credit.

How many pupils were placed in each category?   (AL)

5. A survey was carried out on a number of strawberry plants and from it the following table was made:

Number of strawberries per plant		Frequency
30	I	1
31	II	2
32	IIII	4
33	HHt	
34	HHt III	
35	HHt HHt	
36	HHt HHt IIII	
37	HHt HHt HHt III	
38	HHt HHt HHt IIII	
39	HHt HHt HHt	
40	HHt HHt	
41	HHt III	
42	HHt	
43	III	
44	II	

Copy and complete the table and from it draw a histogram and frequency polygon.

Of what type of distribution might this be an example?

What is the mean?

What is the mode? (EA)

6. The male population of England and Wales in a certain year was distributed amongst the given age groups as follows in the incomplete table.

Age group	Interval (years)	Population (millions)	Height of blocks
0–4	5	2·0	2 units
5–9	5	1·5	1·5 units
10–14	5	1·4	
15–24	10	3·1	
25–34	10	3·3	
35–54	20	6·0	
55–64	10	2·0	
65–74	10	1·4	
75–79	5	0·3	
80–84	5	0·2	
85 and over		0·1	

a. Decide what you must do about the variation in class intervals.

b. Complete the last column.

c. Draw a histogram of the distribution. (SE)

12 Cumulative frequency and standard deviation

In this last chapter, we shall study two topics which unite many of the ideas we have already met in different parts of the book. The two topics are CUMULATIVE FREQUENCY CURVES which develop from the ideas of frequency curves and running totals, and STANDARD DEVIATION which develops from the ideas of deviation from the mean and range.

These two topics are of great practical value, and are used frequently in real life whenever the results of actual investigations have to be analysed.

Cumulative frequency curves

Imagine that you are the Mathematics Examiner for this year's examination, and suppose that the marks scored by 1000 candidates are as shown in the table on the next page.

The table shows that, for example, 42 marks were scored by 48 candidates, and 56 marks were scored by 26 candidates. The righthand column shows how many candidates had a mark in, say, the 30s (157) or in the 70s (47), and so on.

Suppose that as Examiner, you have been told by the Examination Board that not more than 60 per cent of the candidates may pass. The pass mark must now be decided. How would you do this?

	0	1	2	3	4	5	6	7	8	9	Total
0					1		1	3	1		6
1	4	1	6	4	15		4	8	1	1	44
2	5	11	14	9	17	17	3	8	4	5	93
3	14	12	13	16	21	6	32	13	6	24	157
4	25	37	48	22	38	26	23	19	21	45	304
5	29	23	17	37	44	19	26	11	12	18	236
6	15	17	11	16	6	7	12	9	3	2	98
7	12	9	1	8	2	4	6		5		47
8	4		3				1			2	10
9	1	1		1	2						5

Tens digit (row label axis)

Your first idea might be to say that if 60 per cent must pass, then the pass mark will be 40. Let us check this idea.

The number of candidates who score 40 or more is:

Mark	Number of candidates
40–49	304
50–59	236
60–69	98
70–79	47
80–89	10
90–99	5
Total	700

This is 70 per cent of 1000 candidates and not the 60 per cent required by the Board; 100 too many candidates would pass.

To simplify decisions such as this we make use of the idea of a running total, as in Chapter 5, the correct name of which is CUMULATIVE FREQUENCY.

First calculate a cumulative frequency table:

Range of mark	Greatest mark in interval	Number of candidates in interval (frequency)	Total number of candidates scoring the greatest mark or less (cumulative frequency)
0–9	9	6	6
10–19	19	44	50
20–29	29	93	143
30–39	39	157	300
40–49	49	304	604
50–59	59	236	840
60–69	69	98	938
70–79	79	47	985
80–89	89	10	995
90–99	99	5	1000

As with the running total, the cumulative frequency column is calculated by adding in the new frequency in each line.

In this table, the 840 for example, means that 840 candidates have scored *59 or less*; the 985 means that 985 candidates have scored *79 or less*: the 985 includes the 840.

We must now carefully plot the figures in this table on a graph. The horizontal scale is the actual mark scored, the vertical scale is the number of candidates scoring up to and including a particular mark.

It is important to remember that the cumulative frequency of say, 840, is plotted against the *greatest mark* in the interval; that is, *840 is plotted against 59*.

Once all the points have been plotted, they are joined up by a smooth curve and we obtain a CUMULATIVE FREQUENCY CURVE. Such a curve has the characteristic shape of the ogive that we saw in Chapter 5. We can join the points up with a smooth curve since many intermediate points do have a meaning. It is reasonable to ask how many candidates scored 45 or less, or to ask what mark is scored by at least 500 candidates, even though these particular values may not have been plotted. We can assume the cumulative frequency curve to be continuous.

We now return to the problem of the mathematics examiner. If only 60 per cent of the candidates may pass, this means that 400 out of the 1000 must fail. From the graph we see that 400 candidates score 43 or less. The pass mark must therefore be taken as 43.

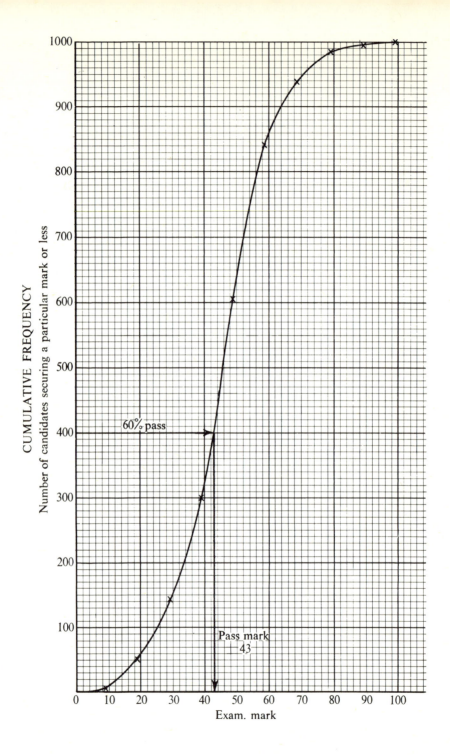

CUMULATIVE FREQUENCY
Number of candidates securing a particular mark or less

60% pass

Pass mark
43

Exam. mark

Percentiles and quartiles

The cumulative frequency curve can be used for many tasks. For example, if we want to find the median score we have only to read off the mark of the 500th candidate. From the graph this comes to 46.

The simplest way to deal with all problems relating to this sort of curve is to divide the vertical scale into 100 equal parts. This means that as well as 'number of candidates', you will also have 'percentages'. In the example we have taken so far, this is particularly easy since there are 1000 candidates.

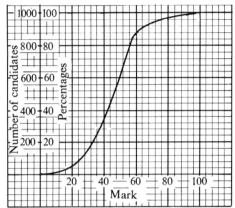

The marks corresponding to the percentages are called the PERCENTILES. The 40th percentile is 43 marks; the 50th percentile is 46 marks.

If 5 per cent of the candidates are to be given a grade 1 pass, we find this also

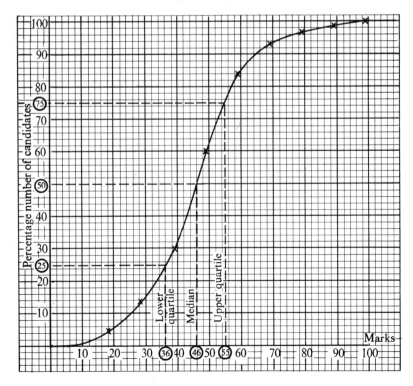

from the cumulative frequency curve. The 95th percentile is 73, so candidates scoring 73 or more are awarded a grade 1.

The 25th, 50th and 75th percentiles are of particular importance. These are called the LOWER QUARTILE, the MEDIAN and the UPPER QUARTILE. These are the marks which divide the total number of candidates into four quarters.

The difference between the two quartile marks is called the INTERQUARTILE RANGE; it is the range of marks which includes exactly one-half of the total population. In this case it is $55 - 36$, that is 19 marks.

Sometimes this range is halved and used as a measure of spread for the whole table. We then have the SEMI-INTERQUARTILE RANGE. In our example,
$$\text{SIQR} = \tfrac{1}{2}(55 - 36) = 9\tfrac{1}{2} \text{ marks.}$$

Exercise A

1. Use the graph on page 110 to answer the following questions:

a. Give the marks corresponding to the 10th percentile, the 32nd percentile and the 80th percentile.

b. What percentile mark is 22, 60, 91?

c. If the top 5 per cent of candidates are given a grade 1 and the next 10 per cent given grade 2, give the range of marks corresponding to these two grades.

2. The ages of the 25 members of the third form are grouped as follows:

Age in y and m	13–13.2	13.3–13.4	13.5–13.6	13.7–13.8	13.9–13.10	13.11–14
Number of children	2	4	6	7	4	2

Design and complete a cumulative frequency table for the age of the class, and draw the corresponding cumulative frequency curve. Mark your vertical scale with both a scale showing numbers of children from 0–25 and a percentage scale from 0 to 100.

From your graph, read off:

a. The median age.

b. The age of the 25th percentile.

c. The age of the 90th percentile.

How many children are there in the class younger than $13\tfrac{1}{4}$ years?

3. The ages of 50 pupils at a primary school are grouped as follows:

Age in years	7–7½	7½–8	8–8½	8½–9	9–9½	9½–10	10–10½	10½–11
Number of children	2	4	8	11	10	3	9	3

Complete a cumulative frequency table, draw a cumulative frequency curve, and from it estimate the median age, the age of the 40th percentile, and the number of children older than 8 years, 7 months.

4. Construct cumulative frequency curves showing: (*a*) weights, (*b*) heights and (*c*) ages for your class. Who has the median weight? The median height? The median age? Who is the 25th and 75th percentile in each case?

5. Find the quartile values for the tables in questions 2 and 3 above. Calculate the interquartile range and the semi-interquartile range in each case.

Standard deviation

Suppose the weekly test scores of two children over a period of five weeks are:

	Elizabeth	*John*
	8	3
	5	9
	6	9
	4	3
	7	6
Totals	30	30

The total score in each case is 30, and the weekly average is 6. However, as we saw in Chapter 8, Elizabeth's range or spread is $8-4 = 4$ marks, while John's range is $9-3 = 6$ marks. We would say that John is more variable than Elizabeth.

Another way of showing this variability was to calculate the mean deviation from the mean, which we also did in Chapter 8, page 84. This would give the *differences from the average mark of 6 as*:

	Elizabeth	*John*
	+2	−3
	−1	+3
	0	+3
	−2	−3
	+1	0
Totals	0	0

If we took into account the plus and minus signs, the totals came to zero. To overcome this difficulty, and give us a measure of variability, we ignored the signs, giving the *mean deviation from the mean as*:

Elizabeth $\qquad 2+1+0+2+1 = \dfrac{6}{5} = 1{\cdot}2$

John $\qquad 3+3+3+3+0 = \dfrac{12}{5} = 2{\cdot}4$

This confirms our impression that John is more variable than Elizabeth.

Another way of avoiding the difficulty of the plus and minus signs is to square all the differences. This means that all the signs will be plus since the square of a negative number is positive. For example:

$$(-3)^2 = (-3) \times (-3) = 9.$$

The following table shows the *squares of the deviations* from the average mark:

Elizabeth		John	
Deviation	Deviation squared	Deviation	Deviation squared
+2	4	−3	9
−1	1	+3	9
0	0	+3	9
−2	4	−3	9
+1	1	0	0
Total	10	Total	36

We now find the average for each:

$$\text{Elizabeth} \quad \frac{10}{5} = 2$$

$$\text{John} \quad \frac{36}{5} = 7 \cdot 2$$

These two numbers are called the VARIANCES for each set of marks.

One final step is to take the square root of the variance; this gives:

STANDARD DEVIATION for Elizabeth $\quad \sqrt{2} = 1 \cdot 41$

STANDARD DEVIATION for John $\quad \sqrt{7 \cdot 2} = 2 \cdot 68$

Once more, our impression that John is more variable than Elizabeth is supported.

We now have three measures of variability; range, mean deviation and standard deviation. Standard deviation has important applications in later work where it has considerable advantages over the other measures.

The separate steps in the calculation may be summarised as follows:

CALCULATION OF THE STANDARD DEVIATION

a. Calculate the mean.
b. Calculate the differences from the mean.
c. Square the differences.

d. Average the squares.
e. Take the square root.

Exercise B

1. Calculate the standard deviations of the following sets of numbers. (Use a slide rule, or square root tables, for the final square root.)

a. 3, 6, 8, 9, 4

b. 2, 2, 6, 10, 10

c. 3, 4, 5, 6, 7, 8, 9

d. 4, 4, 6, 6, 8, 8

e. 4, 6, 6, 8

f. 4, 4, 6, 8, 8

2. Arrange the sets in question 1, in order of increasing variability, using the standard deviation as the measure of variability.

Applications of the standard deviation

Suppose a class of ten boys gain the following marks in two tests held on successive weeks:

Boy	A	B	C	D	E	F	G	H	I	J
Test 1	4	7	3	9	7	3	11	2	9	5
Test 2	Absent	16	14	28	24	12	26	18	24	18

We want to work out what mark to give A for his second test. It would not be right to given him the same as for test 1, since everyone has a higher mark in test 2 than in test 1.

Let us first calculate the mean score in each test, and the difference from the mean for each boy.

Boy	A	B	C	D	E	F	G	H	I	J	Total	Mean
Mark	4	7	3	9	7	3	11	2	9	5	60	6
Difference from mean	−2	+1	−3	+3	+1	−3	+5	−4	+3	−1		
Mark	Abs.	16	14	28	24	12	26	18	24	18	180	20
Difference from mean	?	−4	−6	+8	+4	−8	+6	−2	+4	−2		

It would not be quite fair to say that since A's score is 2 below the mean in test 1, it should be 2 below the mean in test 2, since the differences in test 2 are nearly all larger than in test 1.

The method that statisticians use to overcome this problem is called STANDARDISING THE DIFFERENCES.

First, calculate the standard deviations for each test:

| | Test 1 | | | Test 2 | |
Mark	Difference from mean	Squared difference	Mark	Difference from mean	Squared difference
4	−2	4	Abs.	—	—
7	+1	1	16	−4	16
3	−3	9	14	−6	36
9	+3	9	28	+8	64
7	+1	1	24	+4	16
3	−3	9	12	−8	64
11	+5	25	26	+6	36
2	−4	16	18	−2	4
9	+3	9	24	+4	16
5	−1	1	18	−2	4
60	0	84	180	0	256

Standard deviations; test 1: $\sqrt{\dfrac{84}{10}} = \sqrt{8\cdot4} = 2\cdot9$

test 2: $\sqrt{\dfrac{256}{9}} = \sqrt{28\cdot4} = 5\cdot3$

Note

a. Difference from the mean columns have been added as a check.

b. The total square differences for test 2 has been divided by 9.

To standardise the difference we now divide each difference by the corresponding standard deviation, this gives the *table of standardised differences.*

	A	B	C	D	E	F	G	H	I	J
Test 1	$-\dfrac{2}{2\cdot9}$	$+\dfrac{1}{2\cdot9}$	$-\dfrac{3}{2\cdot9}$	$+\dfrac{3}{2\cdot9}$	$+\dfrac{1}{2\cdot9}$	$-\dfrac{3}{2\cdot9}$	$+\dfrac{5}{2\cdot9}$	$-\dfrac{4}{2\cdot9}$	$+\dfrac{3}{2\cdot9}$	$-\dfrac{1}{2\cdot9}$
	$=-0\cdot69$	$=+0\cdot34$	$=-1\cdot03$	$=+1\cdot03$	$=+0\cdot34$	$=-1\cdot03$	$=+1\cdot72$	$=-1\cdot38$	$=+1\cdot03$	$=-0\cdot34$
Test 2	?	$-\dfrac{4}{5\cdot3}$	$-\dfrac{6}{5\cdot3}$	$+\dfrac{8}{5\cdot3}$	$+\dfrac{4}{5\cdot3}$	$-\dfrac{8}{5\cdot3}$	$+\dfrac{6}{5\cdot3}$	$-\dfrac{2}{5\cdot3}$	$+\dfrac{4}{5\cdot3}$	$-\dfrac{2}{5\cdot3}$
		$=-0\cdot75$	$=-1\cdot13$	$=+1\cdot51$	$=+0\cdot75$	$=-1\cdot51$	$=+1\cdot13$	$=-0\cdot38$	$=+0\cdot75$	$=-0\cdot38$

A study of this table shows that there is a closer connection between the standardised differences for each boy than there is between the ordinary differences.

The best we can do for A is to assume that his standardised difference for the second test will be the same as that in the first test, that is, −0·69. This means that he will have an actual difference of $-0\cdot69 \times 5\cdot3 = -3\cdot66$.

We shall therefore give A a corrected score of $20 - 3\cdot66 = 16$ to the nearest mark for test 2.

Example

A pupil scores 54 marks in a test, the standard deviation of which is 3 and the mean score is 48. In the next test he was absent, and the test had a standard deviation of 4 with a mean of 62. Decide on an estimated score for this pupil in the second test.

Put the data in a table:

	Pupil's score	*Mean of test*	*Standard deviation of test*
Test 1	54	48	3
Test 2	Absent	62	4

Difference from the mean in test 1: $54 - 48 = +6$

Standardised difference: $+\frac{6}{3} = +2$

Assume that the same standardised difference is obtained in test 2; this gives an actual difference of $+2 \times 4 = +8$.

Estimated score for test 2: $62 + 8 = 70$.

Exercise C

1. Calculate the mean and the standard deviation for the set of numbers 1, 3, 6, 6, 9, 10, 9, 4.

For each number, calculate the difference from the mean, and standardise each difference.

2a. Calculate the standard deviations for the following set of results obtained from a test marked out of ten: 3, 4, 5, 6, 7.

b. A pupil scored 49 in a test in which the mean score was 65 and the standard deviation 8. In another test, the mean score was 52 and the standard deviation was 6. What score should the same pupil have obtained in the second test for it to be equivalent to his score in the first test? (AL)

3a. Explain the difference between the two statements:

'Brown's mark in mathematics is 75 per cent.'

'Brown's mark in mathematics is at the 75th percentile.'

b. The following table gives the distribution of marks gained by a group of pupils in a test for which the maximum mark is 9:

Mark	1	2	3	4	5	6	7	8	9
Frequency	1	0	6	15	26	22	16	9	5

i. How many pupils are there in the group?

ii. What is the range of this distribution?

iii. Calculate the mean mark of the distribution.

iv. Write a sentence saying under what conditions the following statement could be fairly reliable:

'On a basis of the given results, we would expect that in another group of 250 pupils about 75 of them would get more than 6 marks.' (EA)

4. The marks obtained by 1500 candidates in an examination were as follows:

Mark	Number of candidates
0–9	20
10–19	85
20–29	155
30–39	230
40–49	360
50–59	235
60–69	175
70–79	130
80–89	95
90–99	15

Plot these results on a cumulative frequency curve.

From your graph, estimate:

a. The number of candidates who achieve the pass mark of 45.

b. The 'commended' mark if 20 per cent of candidates are commended.

13 Exercises

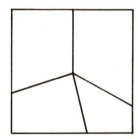

1. Why are pie charts always round? Why are square, or even other shapes not used?

2. In a certain school the pupils are placed into one of five houses. The number of boys and girls allocated to each house is given in the table below.

Name of house	Number of boys	Number of girls
Chester	56	42
Lancaster	48	52
Durham	50	48
York	42	56
Shrewsbury	46	44

Illustrate this information on a clearly labelled bar graph which distinctly shows the number of boys and girls in each house. (AL)

3. Find out the prices of six well-known makes of saloon car and present the information in an attractive manner.

4. Describe any statistical investigation which you made while at school in the last year. You should state clearly:

a. The object of the investigation.

b. How you obtained the information.

c. How you avoided bias.

d. How you tabulated the results. (EA)

5. Use the information given in the following tables A, B and C to answer (*a*), (*b*) and (*c*) below.

Table A. The way in which 6 hours of broadcasting was shared out among various types of programme.

Type of programme	Total number of hours devoted to it
Music	$2\frac{1}{2}$
Plays	$1\frac{1}{2}$
News	$\frac{3}{4}$
Variety	$1\frac{1}{4}$

Table B. The age distribution of 2000 people in a certain area.

Age	0–19	20–29	30–39	40–49	50–59	60–69	70–79
Frequency	720	270	240	320	210	150	90

Table C. The main methods used by 100 children for getting to school.

Method of getting to school	Number of children
Bus	40
Car	10
Cycle	30
Train	5
Walking	15

Exercises

a. Draw an accurate pie chart, using a circle of radius 6 cm to illustrate the figures in table A.

b. Draw a histogram to illustrate the figures in table B.

c. Draw pictograms to illustrate the figures in table C.   (MX)

6. Imagine that you and some of your class have been asked to investigate the way in which pupils at your school spend their leisure hours. You decide to use a questionnaire method of approach and another method of collecting data as well, merely as a check on the results obtained from the questionnaire. Adding any critical comments you may wish to make, briefly describe:

a. The initial planning which would be involved in this investigation, including the drafting of the questionnaire.

b. The details of the second method of collecting data which it is decided Exercises to use.

c. The ways in which the results would probably be analysed and tabulated.

d. An outline of the important points which should be covered in your report of the investigation.   (MX)

7. Illustrate each of these sets of figures with a suitable diagram and give brief reasons for your choice of diagram.

a. The road deaths in a certain city, of young people under the age of 21 years, *according to types of road users.*

Pedestrians	45%
Motor cyclists	25%
Pedal cyclists	15%
Passengers	10%
Drivers (car)	5%

b. The road deaths of young people under the age of 21 years, in a certain city, *throughout one year.*

Jan.	Feb.	Mar.	Apr.	May	June	July	Aug.	Sept.	Oct.	Nov.	Dec.
16	14	10	14	9	20	26	32	24	15	12	18

c. The road deaths of young people, under the age of 21 years, in a certain city, *according to age groups and sex.*

Age (years)	Girls	Boys	Total
$1-4\frac{1}{2}$	9	9	(18)
Over $4\frac{1}{2}-7\frac{1}{2}$	16	8	(24)
Over $7\frac{1}{2}-11\frac{1}{2}$	22	16	(38)
Over $11\frac{1}{2}-13$	15	10	(25)
14–17	48	8	(56)
18–21	35	14	(49)

(SE)

8. The marks obtained by pupils in an examination were as follows:

```
24  82  71  59  32  33  29  20  24  73  59  41  34  31  21
26  64  83  72  13  58  64  39  18  24   1  97  82  73  36
55  20  71  46  71  45  44  19  94   3  16  61  72  83   4
26  43  41  71  64  63  84  19  28  59  51  53  54  90  91
49  64  63  60  39  36  38  47  82  74  76  64  66  54  48
77  45  62  65  51  54  63  72  53  39
```

a. Tabulate these as a frequency distribution, using class intervals 0–9, 10–19, 20–29, etc.

b. Use this to draw a histogram.

c. Use the histogram to find the modal class. (EA)

9a. In an examination each of the candidates paid £1·50 entry fee, from which:

20 per cent went in administrative costs,

30 per cent went in printing costs,

25 per cent went in marking fees,

10 per cent went in school expenses, and the rest went in certificates and prizes.

Draw a pie chart, using a circle of radius 5 cm, to illustrate this distribution.

b. The marks out of 100 scored by a group of 60 children were as follows:

0–9	10–19	20–29	30–39	40–49	50–59	60–69	70–79	80–89	90–100
0	3	5	7	8	14	8	7	5	3

Draw a frequency diagram to represent this data. (Use graph paper.)

c. Using these marks:

i. Determine the modal score—answer as a class interval.

ii. Which class interval includes the pass mark if 75 per cent of the candidates passed?

iii. Which class interval includes the credit mark, if 25 per cent of the candidates obtained a credit pass? (SR)

10. Fare per passenger   Number of tickets

Fare per passenger	Number of tickets
3p	24
4p	34
5p	58
6p	79
7p	198
8p	240
9p	232
10p	102
11p	21
12p	12

The table shows the distribution of bus tickets bought by 1000 people. Find:

a. The average fare paid per passenger.

b. The median fare.

c. The fare paid by the *modal group* of passengers. (SE)

11. The following were the weekly earnings of twelve school leavers on starting work:

£3·75; £4·25; £4·37$\frac{1}{2}$; £4·50; £4·50; £4·50; £4·60; £4·60; £4·75; £5·00; £5·50; £5·62$\frac{1}{2}$.

For those earnings what is:

a. the range,

b. the semi-interquartile range,

c. the mode,

d. the mean? (AL)

12. Illustrate these two tables of statistical information in the most appropriate way.

a. The temperatures recorded on a thermometer throughout one day:

Time	6 a.m.	8 a.m.	10 a.m.	12 a.m.	2 p.m.	4 p.m.	6 p.m.	8 p.m.
Temp. °C	10	12	16	19	21	17	14	12

b. The production of electricity in Western Europe in a certain year. The figures represent millions of units of electricity:

Method of production	Amount
Coal	341
Oil	34
Gas	103
Hydro-electric	122

(SE)

13. Give examples of histograms or frequency polygons of the following types, making a sketch of each, and stating one situation in each case which would lead to such distributions:

a. normal;

b. positive skew;

c. U-shaped;

d. rectangular. (EA)

14. In Chapter 11, page 98, you were asked why a histogram of a distribution with unequal class intervals could not be regarded as entirely accurate. Help to explain your answer by giving all the possible situations which could be represented by this set of data:

Composition of a tutorial group by age	
Age in years	Number of pupils
13	5
14	6
15	5
16–18	3

15. The heights of 100 girls are given below, correct to the nearest cm. Make a frequency table indicating the number of girls of each possible height.

124	120	122	122	124	122	122	124	128	126
128	114	114	120	122	112	124	122	118	118
132	122	120	120	126	124	122	122	122	110
126	124	118	122	132	122	118	130	122	118
130	122	124	122	118	128	116	116	122	126
120	128	110	126	120	128	116	126	124	118
134	120	120	120	122	126	132	124	118	112
120	122	132	124	114	126	116	122	119	119
114	120	122	126	124	120	119	119	130	118
128	122	128	120	120	124	122	118	112	130

Find the mean, median and mode of these heights. Assuming that you have carried out your calculations correctly, comment critically on the degree of accuracy of the results. (MX)

16. The histogram on page 125 opposite shows the number of children arriving at school one morning: thus, five pupils arrived between 8.20 a.m. and 8.25 a.m. There were 209 pupils altogether in the school, though some of them were absent at the time. School starts at 9 a.m.

    a. How many pupils arrived between 8.45 a.m. and 8.50 a.m.?
        30, 105, 35, 847, none of these.

    b. How many pupils arrived between 8.25 a.m. and 8.40 a.m.?
        60, 10, 15, 30, none of these.

    c. How many children were absent that morning?
        4, 14, 109, 18, 9.

    d. What percentage of the children at school that day arrived late?
        20, $12\frac{1}{2}$, 25, 5, 15.

    e. At what time in the morning had the first 125 pupils arrived? One of the following times will be within a minute or two of the correct answer
        8.48 a.m., 8.40 a.m., 9.02 a.m., 8.59 a.m., 8.53 a.m. (EA)

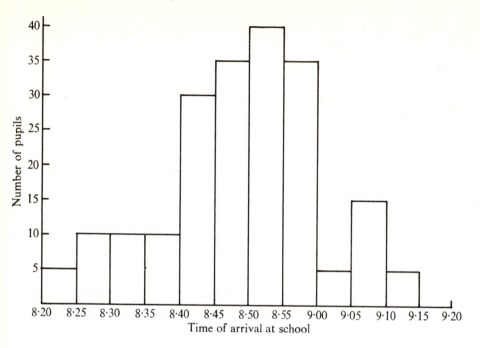

Histogram to show time of arrival at school

Number of pupils

Time of arrival at school

17. Ten candidates received the following marks in two test papers:

Mathematics	49	42	55	51	67	65	73	69	79	75
Spanish	64	54	55	51	51	48	47	44	41	39

a. Plot the points on a scatter diagram and insert the line of best fit.
b. State the general purpose of the scatter diagram.
c. What conclusion can you draw from your diagram?   (SE)

18. The marks awarded to 10 pupils in two mathematics examination papers were as follows:

Pupil	A	B	C	D	E	F	G	H	J	K
Paper I	67	85	46	64	73	52	79	61	76	58
Paper II	58	84	32	57	68	42	76	52	74	47

Plot a scatter diagram and draw a line of best fit.
A pupil P was awarded a mark of 82 on Paper I, but was absent for Paper II.
A pupil Q was awarded a mark of 44 for Paper II, but was absent for Paper I.

Use your scatter diagram and line of best fit to estimate:

    a. The mark that P might have obtained for Paper II.

    b. The mark that Q might have obtained for Paper I.

Clearly indicate on your diagram how you obtained each answer.   (MX)

    19. The following statistics are noted about 10 men:

Annual income in £	800	2000	1200	600	3000	1600	600	1400	1000	2500
Girth (cm)	89	96	81	81	114	99	76	89	81	104
Size of hat	$7\frac{3}{8}$	$7\frac{1}{2}$	$6\frac{3}{8}$	$7\frac{1}{2}$	$6\frac{3}{4}$	7	$6\frac{3}{4}$	$7\frac{3}{8}$	7	7

    Draw two scatter diagrams, one to show how girth varies with the annual income, the other to show how the size of hat varies with the annual income. State briefly what conclusions may be drawn from these diagrams and state also whether any general conclusions may be drawn about the relations between annual income, girth and hat size for the population as a whole.   (EA)

    20. Write down the first 20 four-digit numbers which come into your head. Examine them to see whether you have shown a preference for any individual digit. Compare your results with others in your class. Is any general statement possible?

    21. The letters on the number plate of a car tell you where it is registered. Usually only the last two letters count of a three-letter index mark. For example, if the plate reads 67 CKE then the important letters are KE; similarly FPL 102 E would give PL as the important letters.

    a. Is it reasonable to assume that the majority of the vehicles using the roads near your school are registered locally?

    b. Carry out a survey of number plate letters to see if you can discover the ones allocated to your area.

    22. It is decided to make a set of pigeon holes for the mail addressed to people at an hotel. Are 26 holes needed, one for each letter of the alphabet, or can some initials be combined? Will some initial letters need more than one hole?

    Describe a survey you can conduct to answer this problem. Would it make any difference where the hotel was?

    23. Carry out experiments to determine if there is any correlation between

    a. A person's height and outstretched arms span.

    b. Length of little finger and length of middle finger.

    24a. If the figures relating to two sets of variables are collected and corresponding pairs of figures are used to plot points on a graph, sketch a distribution of the points which would be expected in each of the following cases:

   i. fairly high positive correlation between the variables;

   ii. zero correlation between the variables;

   iii. a correlation coefficient of $+1$ between the variables.

   b. In each of the following cases give one example of a pair of quantities which have:

   i. positive correlation;

   ii. negative correlation;

   iii. zero correlation;

   iv. a correlation coefficient of $+1$.   (AL)

   25a. What is the probability that:

   i. A card picked at random from a pack of playing cards will be an ace?

   ii. A throw of a single die will give a score greater than 3?

   iii. Three successive tosses of a coin will be heads?

   b. How would you calculate the probability that a name picked at random from a class register in a mixed school will be a girl's?   (AL)

   26a. In a class there are 18 boys and 12 girls. The names of all the class members are written on cards which are placed in a box and mixed thoroughly. The cards are then drawn out one at a time and replaced each time. What is the probability that:

   i. The first name out will be a girl's?

   ii. The first name out will be a boy's?

   iii. The first two names out will be girls'?

   b. What is the probability of getting a total of 5 when throwing a pair of dice?   (AL)

   27. Each of two football matches X and Y can result in a home win (H), an away win (A) or a draw (D), and it may be assumed that each result is equally likely. The following table gives the possible results of the two matches:

Match X	H	H	H	A	A	A	D	D	D
Match Y	H	A	D	H	A	D	H	A	D

   a. Copy and complete the following:

Number of draws	0	1	2
Number of results	4		

   b. What is the probability of both matches ending in a draw?

c. What is the probability of there being not more than one draw?

d. How many possible results are there of three matches?

e. What is the probability of all three matches ending in a draw?　(EA)

28. In an examination, 30 candidates each took two mathematics papers, Paper I and Paper II. The results for each candidate were as follows:

Paper I	28	32	53	48	18	79	22	70	52	37	33	50	59	42	68
Paper II	40	55	72	49	31	87	25	75	35	43	42	50	51	66	71

Paper I	35	34	42	42	61	42	41	53	53	49	71	32	47	29	43
Paper II	39	51	46	50	62	40	39	58	65	54	65	43	54	27	45

a. Plot these results on a scatter diagram.

b. Is there any correlation between the results on Paper I and Paper II? (SE)

29. When three coins are tossed, the possible outcomes are:

3 HEADS	2 HEADS, 1 TAIL	1 HEAD, 2 TAILS	3 TAILS
H H H	H H T H T H T H H	H T T T H T T T H	T T T
1 way	3 ways	3 ways	1 way

What are the possible outcomes after tossing:

i. Two coins.　　*ii.* Four coins?

In how many different ways can you get 2 HEADS and 3 TAILS when tossing five pennies?

30. Each member of a class of 30 was told, in turn, to count the number of beans contained in a bag. The set of results was as follows:

```
253  254  248  251  254  250
247  250  251  245  250  253
254  247  303  252  256  255
253  252  249  253  298  252
255  254  254  248  246  254
```

a. Which of these results can be rejected as wildly inaccurate? Use an assumed mean of 250 to determine the best value for the number of beans in the bag, to the nearest whole number.

b. All answers that are not more than 3 from this best value are said to be accurate.

What percentage of the class are accurate counters?

31. The heights of a group of children measured in centimetres to the nearest centimetre were as follows:

    112   114   118   122   124

Calculate the standard deviation for these measurements.

If two more children of heights 110 cm and 126 cm respectively joined the group, without carrying out any further calculations state how you consider (*a*) the mean, (*b*) the standard deviation would be affected giving brief reasons for your answers.   (AL)

32. The figures below show the attendances, in thousands, at an exhibition during the days of two consecutive weeks.

	Sun.	*Mon.*	*Tues.*	*Wed.*	*Thurs.*	*Fri.*	*Sat.*
1st week	29	64	30	53	56	62	65
2nd week	33	66	32	50	57	65	69

a. Calculate and tabulate the seven-day moving averages, expressing each correct to the nearest thousand.

b. Draw a graph to illustrate the data in the table above and superimpose on it a graph of the seven-day moving averages.

c. Describe briefly the trend indicated by the moving-averages graph, giving possible reasons for the trend.   (MX)

Answers

Chapter 1

Exercise B, page 6
1. Total 36; Sectors 160°, 90°, 60°, 50°.
2. Total 18; Sectors 120°, 160°, 60°, 20°.
3. Total 72; Sectors 85°, 125°, 75°, 50°, 25°.

Exercise C, page 11

1.

	A	B	C	D	E	F	G	Total
	17	3	7	5	2	15	11	60
Sectors	102°	18°	42°	30°	12°	90°	66°	360°

2. 24 families.
3. Total 60; Sectors 18°, 150°, 54°, 90°, 48°.
4. Total 30; Sectors 144°, 36°, 24°, 96°, 60°.
6. 3 blonde, 8 auburn, 11 black, 14 brown.
7. Total 180 hours.
a. 38°, 18°, 8°, 6°, 18°, 36°, 22°, 34°, 40°, 52°, 24°, 24°, 40°.
b. Suggested grouping (the figures represent the number of hours):

Sport	19	Feature films ⎫	
Westerns	9	Plays ⎬ 63	
Adventure ⎫ 7		Serials ⎭	
Travel ⎭		Education ⎫ 24	
Variety ⎫ 38		News ⎭	
Pop ⎭		Quiz games	9
		Science fiction	11

9. Although there is a vertical scale of numbers, it does not say what they represent. *Answers*

Chapter 2

Exercise A, page 16
4. Mode is 1.   5. Mode is 2.   6. There must be an equal number of every item.
7. 150 cm is the most common height in the group.

Exercise B, page 18
3. Median 5.   4. Median $6\frac{1}{2}$.   5. Median 65.   6. Median 6.
7. Medians: test 1, 5; test 2, 7; test 3, $5\frac{1}{2}$.

Exercise C, page 20
1. Mean 5·6.   2. Mean 42.   3. New average 35.   5. 45 km/h.
6. 220 km.   7. Total 86.   8. Average 11·6.   9. 6·83.

Exercise D, page 22

3.

	Score	Diff. from mean	
		+	−
	27		12
	29		10
	46	7	
	18		21
	54	15	
	60	21	
Totals	234	43	43
Mean	39		

4.

	Score	Diff. from mean	
		+	−
	8·6	2·9	
	4·2		1·5
	2·9		2·8
	5·8	0·1	
	2·3		3·4
	8·0	2·3	
	7·4	1·7	
	6·4	0·7	
Totals	45·6	7·7	7·7
Mean	5·7		

5. Between 104 and 172.

Exercise E, page 24
1. Mean $45\frac{3}{4}$.   2. Mean $45\frac{3}{4}$.
3. It does not matter.   4. Mean 107·2.
5. Mean 6000.   6. Mean 4·065.

7. Leave the 6·30 value out as being inaccurate.   Mean $= \dfrac{36·54}{9} = 4·06$.

Exercise F, page 25
1. 9·4.   2a. 56; b. 61.   3a. $22\frac{1}{7}$; b. $44\frac{2}{7}$.
4. Mode 43·29; median 43·30; mean 43·30 (to nearest milligramme); mean is best value.   5. Median 10; mean 14.   6. Mode, 2 children.

9a. Include everybody. Mean wage $= \dfrac{24,400}{18} = £1355\frac{5}{9}$.

b. Exclude the owner and the manager. Mean wage $= \dfrac{10,400}{16} = £650$.

10a. 600 km; b. $2\frac{1}{2}$ h.   11a. 2 h; b. 5 h; c. $64\frac{2}{7}$ km/h.
12. Mean $126\frac{2}{3}$; mode is 124.   13a, Mode is light brown.
15. Modes 11·5 and 11·7; median 11·6; mean 11·6 to 3 sig. figs.

16.	Score	Diff. from mean	
		+	−
	17		28·2
	56	10·8	
	45		0·2
	68	22·8	
	20		25·2
	46	0·8	
	33		12·2
	51	5·8	
	109	63·8	
	7		38·2
Totals	452	104·0	104·0
Mean	45·2		

Chapter 3

Exercise A, page 31
2. Mean marks; French 52·6, German 47·9.
a. Estimated French mark 27–30.
b. Estimated German mark 59–61.
3. Mean value of $x = 2·5$; mean value of $y = 3·2$.
When x is 1·9, take y as 2·9.
When y is 5·0, take x as 5·9.
4. Mean of term's work $= 59·2$: mean exam mark $= 53·0$. Estimated examination mark $= 54$.
5. Mean value of $x = 4\frac{7}{9}$. Mean value of $y = 4\frac{1}{9}$.
6. Ignore Janette's mark.

Exercise B, page 37
3. (*i*) Direct; (*ii*) inverse; (*iii*) no correlation.
7. Points all lie on a straight line.
8. Mean discus $= 19·5$ m; mean 100 m $= 12·49$ s.

9. *French*	28	34	36	50	57	68	77	85	86	88
Geography	78	27	48	90	45	69	28	60	91	40

a. No correlation; *b.* no estimate can be given.
10a. −; *b.* +; *c.* 0; *d.* +; *e.* +; *f.* −.
Part (*e*) has a correlation coefficient of +1.
11. Mean load $= 80$ g, mean length $= 20$ cm.
a. load $= 12$ g, *b.* length $= 22·4$ cm.

12a. A and E; *b. F; c, d, e* see diagram.

Chapter 4

Exercise B, page 48
1. a with (iv); *b* with (ix); *c* with (i);
d with (v); *e* with (viii); *f* with (iii);
g with (ii); *h* with (vi); *i* with (vii).
4b. Curves (i), (iii), (v), (vii) and (viii).

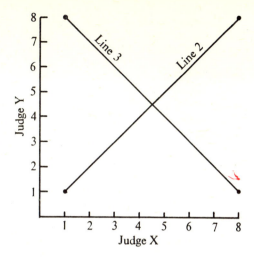

Chapter 5

Exercise A, page 54

1.	*Number*	*Running totals*
	127	127
	31	158
	438	596
	226	822
	407	1229

2.	*Number*	*Running totals*
	11·6	11·6
	4·7	16·3
	18·2	34·5
	9·48	43·98
	15·5	59·48
	10·09	69·57

3.			*Running totals*
	Monday	£1·25	£1·25
	Tuesday	£1·35	£2·60
	Wednesday	£1·75	£4·35
	Thursday	£1·60	£5·95
	Friday	£0·75	£6·70

4.			*Running totals*
	London–Croydon	16 km	16
	Croydon–Redhill	18 km	34
	Redhill–Crawley	16 km	50
	Crawley–Brighton	37 km	87

5a.			*Running totals*
	Monday	186	186
	Tuesday	194	380
	Wednesday	190	570
	Thursday	182	752
	Friday	201	953

5b. 570 served by Wednesday.

7. $\frac{1}{4}$ of the tickets sold on Wednesday of week 1.
$\frac{1}{2}$ of the tickets sold on Thursday of week 1.

Exercise B, page 59
4. A straight line.
5b. Running totals 1, 3, 7, 15, 28, 38, 45, 48, 49.
c. Running totals 1, 5, 11, 23, 43, 88.

d. Running totals 20, 31, 34, 36, 37, 38.
e. Running totals 25, 35, 39, 42, 44, 49, 58, 80.
6. Scores were 17, 43, 29, 12.

Chapter 6

Experiment 2. Expected probability of throwing a 5 is $\frac{1}{6}$.
Expected probability of throwing an even number is $\frac{1}{2}$.
Experiment 3. *a.* $\frac{1}{52}$; *b.* $\frac{1}{4}$; *c.* $\frac{1}{13}$; *d.* $\frac{3}{13}$; *e.* $\frac{1}{2}$.
Experiment 4. Birthday on the same day, $\frac{1}{365}$ (or $\frac{1}{366}$).
Birthday in the same month, $\frac{1}{12}$ (or $\frac{31}{365}, \frac{30}{365}, \frac{28}{365}$ or $\frac{31}{366}, \frac{30}{366}, \frac{29}{366}$)
Experiment 5. $\frac{3}{7}$.

Exercises, page 68
1. Lonely Lady, $\frac{1}{6}$; Just Janie, $\frac{1}{11}$; Pretty Polly, $\frac{2}{27}$: Kay's Delight, $\frac{1}{2}$;
On the Beach, $\frac{1}{34}$; Veronica, $\frac{2}{3}$.

2.

				Totals			
	6	7	8	9	10	11	12
	5	6	7	8	9	10	11
	4	5	6	7	8	9	10
Red die	3	4	5	6	7	8	9
	2	3	4	5	6	7	8
	1	2	3	4	5	6	7
		1	2	3	4	5	6

Blue die

Score of 2, $\frac{1}{36}$;
Score of 7, $\frac{6}{36} = \frac{1}{6}$;
Score of 11, $\frac{2}{36} = \frac{1}{18}$.

3.

5p		
Head	(H, H)	(T, H)
Tail	(H, T)	(T, T)
	Head	Tail

1p

Two heads $\frac{1}{4}$; Penny H, fivepenny T, $\frac{1}{4}$;
One H, one T, $\frac{1}{2}$.

4. Eight possible outcomes; all three coins show a tail, $\frac{1}{8}$.
5. *a.* (i) $\frac{1}{6}$, (ii) $\frac{1}{3}$, (iii) $\frac{1}{36}$; *b.* (i) $\frac{1}{4}$, (ii) $\frac{4}{13}$.

Chapter 8
Exercises, page 85

1.

	Mean	Range	Mean deviation from the mean
Joan	8·3	5	1·80
Margaret	7·9	2	0·36
Ann	3·9	2	0·54
Helen	6·7	2	0·70

Joan has the best total marks but is also the most variable.

2.		Mean	Mean deviation
	A	10·1	1·92
	B	9·8	2·44
	C	10·0	2·40

A is the most reliable.

3.		Mean	Range	Mean deviation
	a.	13	18	5·0
	b.	13	18	5·4
	c.	13	18	9·0
	d.	13	18	2·6
	e.	13	18	6·2

Chapter 9
Exercises, page 89

1. Continuous.
2. Discrete. Possibly during stages 1 or 4, but this cannot be certain; number of passenger miles is *at least*
$$15 \times \tfrac{1}{2} + 12 \times 1 + 7 \times 1 + 8 \times 1 + 10 \times 1 + 12 \times \tfrac{1}{2} \backsimeq 50;$$
total fare is *at least* 64 units = £1·92.
3. High water at 11.40 a.m. Ship can enter between 8.15 a.m. and nearly 3.0 p.m.
4. Transplant after 2 weeks and 2 days.
5. Tenth penny spent on Sunday.

Chapter 10
Exercises, page 95

1.		Simple average	Weighted average
	A	$47\frac{5}{6}$	46·1
	B	43	43·7
	C	$54\frac{2}{3}$	54·4
	D	$54\frac{5}{6}$	53·8

2. Alison, $59\frac{12}{13}$; Ben, $45\frac{6}{13}$.
3. *a.* 69; *b.* $70\frac{1}{3}$.
5. Moving averages $782\frac{1}{4}$, $801\frac{1}{4}$, $819\frac{1}{2}$, 835, 854, $875\frac{1}{2}$, $899\frac{1}{4}$, $918\frac{1}{2}$, $932\frac{3}{4}$.

6. Month	£	3-monthly moving totals	3-monthly moving average
January	150		
February	180	530	$176\frac{2}{3}$
March	200	570	190
April	190	695	$231\frac{2}{3}$
May	305	815	$271\frac{2}{3}$
June	320	865	$288\frac{1}{3}$
July	240	825	275
August	265	785	$261\frac{2}{3}$
September	280	860	$286\frac{2}{3}$
October	315	885	295
November	290	870	290
December	265		

Chapter 11

Exercise, page 105

3. Class interval	Frequency
50–54	2
55–59	7
60–64	1
65–69	4
70–74	2
75–79	3
80–84	2
85–89	0
90–94	1

4. Class interval	Frequency
11–15	1
16–20	2
21–25	3
26–30	6
31–35	9
36–40	6
41–45	0
46–50	3
51–55	6
56–60	3
61–65	1

21 failed, 15 passed,
4 passed with credit.

5. Number of strawberries per plant	Frequency
30	1
31	2
32	4
33	5
34	8
35	10
36	14
37	18
38	19
39	15
40	10
41	8
42	5
43	3
44	2

A bell-shaped or normal distribution; mean $= 37\frac{51}{124}$; mode $= 38$.

6. *Height of blocks:* 2·0, 1·5, 1·4, 1·55, 1·65, 1·5, 1·0, 0·7, 0·3, 0·2. Last group: 85 and over, not included in histogram.

Chapter 12

Exercise A, page 112

1a. 25, 40, 57; *b*. 7th, 85th, 100th; *c*. grade 1, 100 − 73; grade 2, 72 − 61.

2. Age from 13 years up to and including value given	Cumulative frequency
13.2	2
13.4	6
13.6	12
13.8	19
13.10	23
14.0	25

3. Ages from 7 years up to and including value given	Cumulative frequency
$7\frac{1}{2}$	2
8	6
$8\frac{1}{2}$	14
9	25
$9\frac{1}{2}$	35
10	38
$10\frac{1}{2}$	47
11	50

a. Median 13 y 6 m; *b*. 25th percentile 13 y 4 m; *c*. 90th percentile 13 y 10 m. 3 children younger than $13\frac{1}{4}$.

Median age $= 9$ yr;
40th percentile $= 8$y 9m;
35 children older than 8y 7m.

5. Number 2: IQR = 4 months; SIQR = 2 months; Quartiles, 13·4, 13·8.
Number 3: IQR = 1·5 years; SIQR = 0·75 years; Quartiles, 8·45, 9·95.

Exercise B, page 115
1a. 2·3; *b.* 3·6; *c.* 2·0; *d.* 1·6; *e.* 1·4; *f.* 1·8;
2. *e, d, f, c, a, b.*

Exercise C, page 117
Mean 6; standard deviation 3.

Number	1	3	6	6	9	10	9	4
Diff. from mean	−5	−3	0	0	+3	+4	+3	−2
Standardised difference	$-\frac{5}{3}$	−1	0	0	+1	$+\frac{4}{3}$	+1	$-\frac{2}{3}$

2a. Standard deviation 1·4; *b.* 40.
3b. (*i*) 100; (*ii*) 8 marks; (*iii*) 5·7.
4a. 800 pass; *b.* 66.

Chapter 13

8.

Class interval	Frequency
0–9	3
10–19	5
20–29	10
30–39	10
40–49	10
50–59	12
60–69	13
70–79	12
80–89	6
90–99	4

Modal class 60–69.

9a. Sectors are, 72°, 108°, 90°, 36°, 54°;
c. (*i*) 50–59, (*ii*) 40–49, (*iii*) 70–79.
10a. 7·761p; *b.* 8p; *c.* 8p.
11a. £1·87½; *b.* 22p; *c.* £4·50; *d.* £4·66.

14. The last group, 16–18, could be:

16	17	18
3	0	0
0	3	0
0	0	3
2	1	0
2	0	1
0	2	1
1	2	0
1	0	2
0	1	2
1	1	1

15.

Height	Frequency	
110	2	
112	3	
114	4	
116	8	
118	10	
120	14	
122	22	
124	12	
126	9	
128	7	mean = 121·84
130	4	median = 122
132	4	mode = 122
134	1	

16a. 35; *b.* 30; *c.* 9; *d.* 12½; *e.* 8.53 a.m.

18a. 80; *b.* 55.

24a. (*i*) A fairly narrow band of points sloping upwards from left to right.

  (*ii*) No band present.

  (*iii*) All points on a straight line sloping upwards from left to right.

25a. (*i*) $\frac{1}{13}$, (*ii*) $\frac{1}{2}$, (*iii*) $\frac{1}{8}$.

26a. (*i*) $\frac{2}{5}$, (*ii*) $\frac{3}{5}$, (*iii*) $\frac{4}{25}$; *b.* $\frac{1}{9}$.

27a.

Number of draws	0	1	2
Number of results	4	4	1

b. $\frac{1}{9}$; *c.* $\frac{8}{9}$; *d.* 27; *e.* $\frac{1}{27}$.

29.

2 heads	1 head, 1 tail		2 tails
HH	HT	TH	TT

(*ii*)

4 heads	3 heads, 1 tail	2 heads, 2 tails	1 head, 3 tails	4 tails
HHHH	HHHT	HHTT	TTTH	TTTT
	HHTH	HTTH	TTHT	
	HTHH	HTHT	THTT	
	THHH	TTHH	HTTT	
		THHT		
		THTH		
1	4	6	4	1

There are 10 ways of getting 2 heads and 3 tails.

30a. Omit 303, 298; best value 251; *b.* 70 per cent.

31. Standard deviation 4·56.

  a. Mean remains unchanged at 118.

  b. Standard deviation increases.

32a. Moving averages are: $51\frac{2}{7}$, $51\frac{6}{7}$, $52\frac{1}{7}$, $52\frac{3}{7}$, 52, $52\frac{1}{7}$, $52\frac{4}{7}$, $53\frac{1}{7}$.